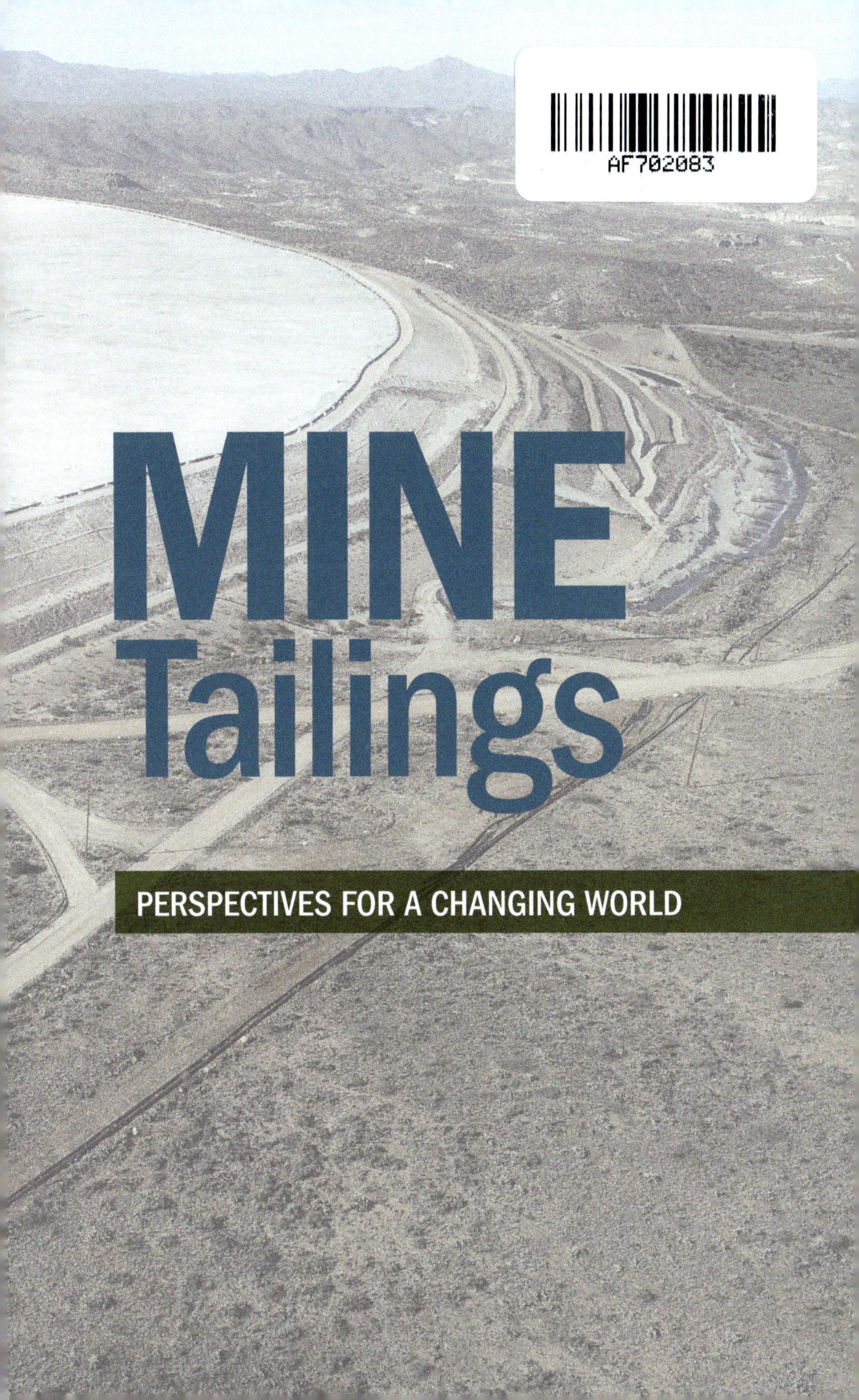

MINE Tailings

PERSPECTIVES FOR A CHANGING WORLD

MINE Tailings

PERSPECTIVES FOR A CHANGING WORLD

Edited by Resa Furey and John F. Lupo

Published by the Society for Mining, Metallurgy & Exploration

Society for Mining, Metallurgy & Exploration (SME)
12999 E. Adam Aircraft Circle
Englewood, Colorado, USA 80112
(303) 948-4200 / (800) 763-3132
www.smenet.org

The Society for Mining, Metallurgy & Exploration (SME) is a professional society whose more than 15,000 members represent all professionals serving the minerals industry in more than 100 countries. SME members include engineers, geologists, metallurgists, educators, students, and researchers. SME advances the worldwide mining and underground construction community through information exchange and professional development.

Information contained in this work has been obtained by SME from sources believed to be reliable. However, neither SME nor its authors and editors guarantee the accuracy or completeness of any information published herein, and neither SME nor its authors and editors shall be responsible for any errors, omissions, or damages arising out of use of this information. This work is published with the understanding that SME and its authors and editors are supplying information but are not attempting to render engineering or other professional services. Any statement or views presented herein are those of individual authors and editors and are not necessarily those of SME. The mention of trade names for commercial products does not imply the approval or endorsement of SME.

ISBN 978-0-87335-488-2
eBook 978-0-87335-489-9

Electronic edition published 2020.

Library of Congress Control Number: 2020949900

Contents

Preface

Nearly every recent article on tailings starts by mentioning a large tailings dam failure. Mount Polley (2014), Fundão (2015), Brumadinho (2019), and other prior failures serve to remind readers what is at stake. The consequences of the recent failures have been so devastating they have pushed conversations about the risks inherent in these structures beyond the mining community into the general population. In fact, society no longer trusts the mining community to deal with this alone and in almost every jurisdiction, is imposing new, more stringent rules regarding governance and in some cases, new technical standards and guidelines. We are left to question how we can best address the risks associated with tailings disposal as we and others now see them, and in so doing, transform the image of the mining industry and perhaps the industry itself.

With this as a backdrop, the Society for Mining, Metallurgy & Exploration (SME) challenged tailings and mining professionals to re-imagine the future of tailings. The Mine Tailings: Perspectives for a Changing World symposium, held at the SME 2020 annual conference, started that conversation. Over three days, tailings professionals from around the world gathered to discuss tailings storage practices and the changes both the industry and the world want and need. The discussions squarely focused on how we, as an industry, can collectively make changes that will eliminate catastrophic tailings dam failures and lead to better outcomes for the industry and society.

This book provides summaries of the presentations and discussions at the symposium's five sessions. The titles of the sessions are as follows:

- Tailings Dam Failures—Lessons Learned from Technical and Governance Realities,
- State of Practice,
- Tailings Governance,
- Tailings Continuum—Selection of the Right Tailings Technology, and
- The Importance of Culture in Safe Tailings Management.

There is an interesting juxtaposition in mining: As an industry, mine risk conversations have historically focused on the business side—ore bodies, production costs, safety—with less focus on risks associated with non-revenue-producing elements such as mine waste. Yet tailings practitioners argue that they focus almost exclusively on risk. A case in point is that year after year, tailings conferences are dedicated to understanding and improving the technical practices associated with tailings. Although tailings risk conversations often begin with a concentration on the human element, they quickly pivot to the technical, for this is where tailings practitioners feel comfortable. Tailings professionals are ultimately engineers and scientists, not social scientists. However, conversations that focus exclusively on the technical miss the point: Technical excellence alone will not stop tailings dam failures. We must focus tailings conversations to include the risks associated with the human element and cultural aspects, which are present in every aspect of tailings management. Recent tailings dam failures are reminders that we have not yet taken these risks seriously enough.

The Mine Tailings: Perspectives for a Changing World symposium was organized to "put the industry's waste issues on the table." This is a "people" problem caused by people. Humans and the environment are the victims of tailings failures. The root of tailings risk sits with us, as time has shown tailings dam failures to be a human issue resulting from our inability to self-assess. This raises questions such as: Are we over-confident in our work? Are we unable to assess risk because of our own personal bias? Through sharing and conversation, the symposium participants recognized risks associated with our approach to tailings management and existing structures and discussed the gaps that need to be addressed, including how the behavior of tailings and mining professionals must change. The human element of risk *must* be recognized so it can be talked about openly, given the attention it deserves, and adequately addressed. We need to *own* this problem and the impact of our actions. We have the power to change this; when we own our actions, we can act differently for a different outcome.

The Mine Tailings: Perspectives for a Changing World symposium started the conversation at SME, and those attending and contributing to the event moved the discussion forward in a meaningful way. We have a long way to go, and SME remains committed to zero harm, including zero tailings facility failures.

—Resa Furey and John F. Lupo, Editors

SESSION 1

Tailings Dam Failures— Lessons Learned from Technical and Governance Realities

CHAIRS

Kimberly (Finke) Morrison, Newmont
Dirk van Zyl, University of British Columbia

PRESENTERS AND PANELISTS

C.D. (Lyn) Anglin, Anglin & Associates
Allen H. Gipson Jr., Consultant
Kalidas Madhavpeddi, Azteca Consulting
Harvey McLeod, Klohn Crippen Berger Ltd.
Dirk van Zyl, University of British Columbia

NOTETAKERS

Katie Kruger, Freeport-McMoRan Inc.
Dennis Rugg, Golder Associates

INTRODUCTION

This session kicked off the Society for Mining, Metallurgy & Exploration (SME) Mine Tailings Symposium at the SME 2020 Annual Conference and Expo and focused on tailings dam failures and lessons learned from technical and governance realities. The session covered topics including a broad look at managing reputational risk; a discussion of systems-related failures with a comparison to other industries; a review of tailings failure case history examples, including the role of human error; a look at the Mount Polley breach five years later with a focus on reputational management; and finally, an overview of key tailings facility failures and lessons learned from a social, regulatory, technical, and governance perspective. Following the presentations, the panel and the audience participated in a guided question-and-answer session.

PANEL DISCUSSION: LESSONS LEARNED FROM TECHNICAL AND GOVERNANCE REALITIES

The panel provided an open dialogue on the lessons learned from tailings dam failures. Panelists delved deeper into the issues the industry faces with respect to root causes of failures and repercussions of failures, including reputational risks. The panel discussion indicated general agreement that there are many ways to apply lessons learned in the past to make for a safer tomorrow, including industry actions, the development of standards and guidelines, and improved implementation of governance.

Within the industry, companies need to manage their own reputation in parallel with the reputation of the industry itself. To do this, a company must be proactive and build relationships before there is an incident, not just when something has gone wrong. This can be achieved by engaging cross-functional teams and sharing our successes along with our failures. An example of this is the Mount Polley failure; the company worked hard to build a reputation prior to the failure, so when the failure occurred, it was easier for the company to secure resources for the cleanup.

As an industry, we need to be open and viewed as a trusted source of information. To build trust, we must practice transparency by sharing information about our successes and failures. We must also be competent in the subject matter by understanding failure modes and how uncertainty may or may not be captured by a factor of safety. One needs to understand the details of a structure to come to a reasonable conclusion on risk. We should think through the process of communicating risk to the public. In the case of Mount Polley, no one was pleased with the failure, but the panel hoped that all were pleased with the subsequent response.

A successful team relies on engagement from the engineer of record (EoR). The EoR has a professional responsibility for human safety and to protect society. Therefore, when laying out the terms of reference for an EoR, expectations should be clear but not overly prescriptive. It is beneficial to engage the EoR starting in the design phase and maintain the relationship through operations. The relationship between the EoR and owner/operator should be based around a culture of protecting human life and environment. The ultimate responsibility for tailings storage facility safety goes beyond the EoR, should be shared across a multi-tiered system made up of individuals with different backgrounds and perspectives, but rests in the hands of the owner/operator.

The governing board and the management team play different roles in tailings management, but most boards are conscious of the risks associated with tailings storage facilities. The board can ask the management team to perform an audit and present the audit to the board. Although this is frequently done, it is not often reported publicly. Boards enhance their own understanding of tailings management by bringing in technical experts. Boards also routinely tour facilities to increase understanding and send a message to management and operations teams that tailings are an important aspect of the overall business.

Guidelines and standards provide a basis for mutual understanding and expectations for tailings management. Guidelines and standards should reflect learnings from failures. They should be well thought out, vetted, and anchored in the true causes of failure, rather than knee-jerk reactions. The goal is to get to the root cause of the failure to prevent future failures. The industry can use the *Global Industry Standard on Tailings Management* (GISTM) to help guide future decisions. The original terms of reference for the GISTM were broad, and the group had to pull together to develop a standard drawing from the varied expertise of the panel.

The GISTM is about flexibility and acceptance rather than enforcement. Trying to implement and enforce the standard worldwide may be too far-reaching and unlikely to be successful. The larger companies are more likely to adopt the standard. If the standard is shown to have value, then implementation will be more successful and smaller institutions may begin to adopt them. Regardless, the standard can be a helpful reference even if governmental agencies or regulations do not directly adopt it.

The panel discussed a potential tailings facility safety certification process in conjunction with the GISTM, and opinions on the topic varied. On one hand, a stepped process to certifying a tailings facility might be easier to incorporate and adopt over time. On the other hand, a tailings facility could be rated as simply "pass" or "fail" from a governance perspective. Additionally, the industry needs to respect and understand that societies have different expectations and regulations. If standards become too prescriptive, support might wane. Certification of tailings facilities also has investment and insurance implications; a certified site that has strongly implemented the GISTM might be more appealing to investors and insurance firms. Financial influences may be the key to getting things done quickly.

A robust auditing system may help to restore public, insurance, and investor confidence in the mining industry. When developing an internal auditing program, consider complacency; fresh eyes may be needed for a new perspective. The industry should continue to foster relationships with regulators. Currently, the panel feels that there is not enough open dialogue. Mining companies should be responsive to discussions regarding their own protection as much as the public's. From a regulatory perspective, we need to entice experienced engineers with a solid understanding of tailings to join the regulatory bodies. Mining companies cannot do it all but can help lead the right people to the right positions, especially when open dialogue exists between mining companies and regulatory bodies.

CLOSING AND VALUE OF LESSONS LEARNED

This session provided varied perspectives on lessons learned from historical failures and recommendations for moving past these failures. Several common themes and topics occurred throughout the presentations and the panel discussion. Open lines of communication and the sharing of information and best practices were highlighted as requirements for success moving forward. The industry and individual companies can improve our collective reputation by being more open and forthcoming with the public. Governance and regulatory oversight can be improved by eliminating information silos and maintaining dialogue between operators and governing bodies (e.g., Mount Polley). The elimination of silos within individual companies or operating mines can improve our technical understanding of the overall tailings systems, thereby minimizing errors and miscommunications that can lead to adverse results. Improved governance and technical guidance, along with updated organizational structures and improved internal auditing systems, can lead to incremental cultural changes throughout our industry that can help improve dam safety. An engaged EoR, the GISTM, transparency, and a potential new and/or improved auditing and certification process can all help to restore public, insurance, and investor confidence in the mining industry.

Integrating Tailings Technical and Governance Realities to Better Appreciate System Failures

Dirk van Zyl
University of British Columbia, Vancouver, British Columbia, Canada

INTRODUCTION

Although there are immediate technical reasons that tailings dams fail, the overall technical and governance system of tailings management is, in most cases, the root cause. This paper explores this matter to evaluate the importance of understanding and managing this system appropriately.

The terms of reference of recent failure investigation panels are narrowly focused on the immediate technical reasons. A broader system review requires that the company's management approaches, design, construction, oversight activities, and decision-making be scrutinized and evaluated. Such an evaluation is time consuming, depending on full transparency of documentation as well as a broad survey of all parties involved. It is doubtful that any corporate lawyer would allow this to happen from a legal perspective within a company. Even a confidential internal investigation can result in legal challenges as the outcomes may be discoverable.

Despite these challenges to identify the system components and their interactions that contributed (or may contribute in the future) to a failure, it is incumbent on the corporate managers of corporate tailings functions to develop a better understanding of the system as well as the potential pitfalls along the way.

UNDERLYING FACTORS OF SYSTEMS FAILURES

In his book *Drift into Failure*, Sidney Dekker (2011) describes five concepts that together may characterize drift to failure:

- **Scarcity and competition.** Organizations operate and make decisions in an environment with real constraints. There can also be strong competitive forces in play. This all leads to decision-making within these constraints.

- **Decrementalism, or small steps.** Small incremental changes may occur. A series of decisions are made that may not seem related, but in the long term, they impact the outcomes. Some of these decisions also rely on incomplete emerging information.
- **Sensitive dependence on initial conditions.** A comprehensive understanding of initial conditions can be critical to the outcome of a series of complex decisions. Regulators are aware of only some of the interdependencies.
- **Unruly technology.** Uncertainty is included in the evaluations; analyses are performed, and decisions are made. However, there are always unknowns that play a role in the "unruly technology."
- **Contribution of protective structure.** Risky technology is surrounded by structures (regulations, permits, etc.) that are meant to keep it safe. Personnel changes can occur, and the knowledge is not transferred along the line of the protective structure.

Because of all these matters Dekker (2011) strongly suggests that we move away from the Isaac Newton and René Descartes ideas of cause and effect. Although it is a challenge to find the "immediate technical cause of a failure" following the Newton–Descartes approach, it is much more difficult to find an answer if one sees it as a complex system. Safety-critical organizations are complex adaptive systems. These organizations must pay attention to diversity. "With diversity, a system has a larger number of perspectives to view a problem with a larger repertoire of possible outcomes" (Dekker 2011).

There clearly are differences in how one investigates complex systems failures versus investigating the immediate technical cause of a failure. The sidebar provides a quote from Dekker (2011) on tackling investigations in such an intricate problem. This is clearly not straightforward, and indeed, engineers by themselves may not be the best suited to evaluate such a system.

Another important insight of Dekker (2011) is the inclusion of diversity to reduce the overall chances for drifting into failure. Diversity impacts the five factors just identified and results in a much more resilient outcome.

Tailings facilities are complex systems. The design, construction, operations, and closure of these facilities pose complicated issues. Over the last two decades, with the Mining Association of Canada (MAC) taking the lead, the industry has adopted tailings governance approaches that are making important contributions to complex system safety management. MAC has developed a *Table of Conformance* to identify elements of both the *Guide*

INVESTIGATING FAILURES IN COMPLEX SYSTEMS

We can all work on letting a post-Newtonian ethic of failure emerge if we embrace systems thinking more seriously than we have before. In a post-Newtonian ethic, there is no longer an obvious relationship between the behavior of parts in the system (or their malfunctioning, for example, "human errors") and system-level outcomes. Instead, system-level behaviors emerge from the multitude of relationships, interdependencies, and interconnections inside the system, but cannot be reduced to those relationships or interconnections. In a post-Newtonian ethic, we resist looking for the "causes" of failure or success. System-level outcomes have no clearly traceable causes as their relationships to effects are neither simple nor linear.

In such an ethic, we acknowledge that our selection of causes (or events or contributory factors) is always an act of construction. Our extraction of accountability is that too. It is the creation of a story, not the reconstruction of a story that was already there. There is no objective way of constructing a story of what happened—all analytical choices (of what we look at, where we look, who we talk and listen to, how we write up what we learn) are affected by our background, preferences, language, experience, biases, beliefs, and purposes. We can at best hope and aim to produce authentic stories, stories that are true to experience, true to the phenomenology of being there, of being suspended inside complexity.

This means there are always multiple possible authentic stories, and we should not seek to affirm or legitimate one interpretation (for example, one story of what happened) and let it drown out all the others.

Source: Dekker 2011

to the Management of Tailings Facilities (MAC 2019b) and *Developing an Operation, Maintenance, and Surveillance Manual for Tailings and Water Management Facilities* (MAC 2019a) to be implemented to be "in conformance" (MAC 2019c).

Using this conformance scheme helps to integrate a complex system and can be seen as a way to manage this complex system using diversity; the conformance table considers more than just the technical aspects. This author considers this a large step forward and sees it also as a challenge not to see this as a milepost and not keep looking for ways to the broader inclusion of

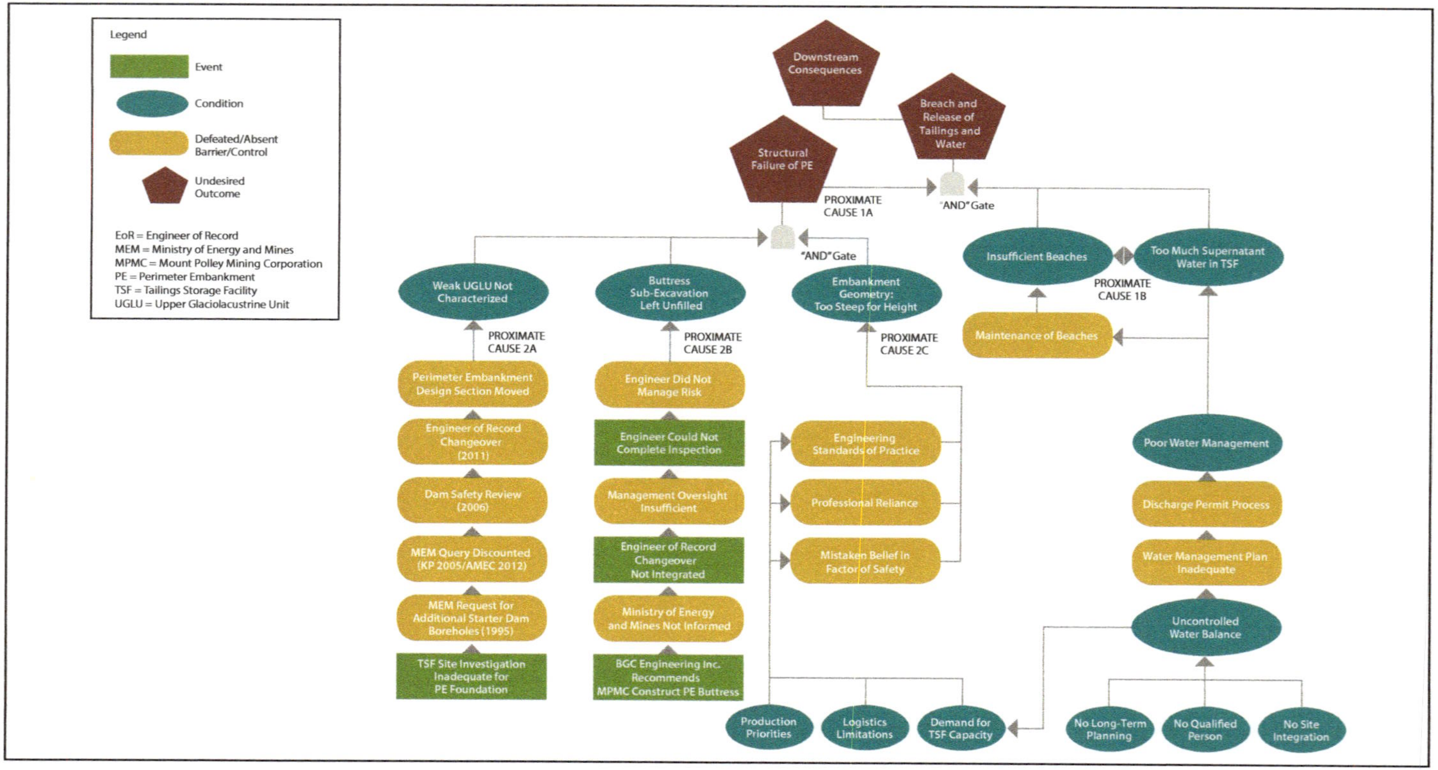

Figure 1 Root cause analysis event and causal factor tree for the failure of the Mount Polley tailings storage facility

diversity in the management of the technical and governance aspects of tailings facilities.

EXAMPLE OF SYSTEMS APPROACH IN EVALUATING A TAILINGS FAILURE

An example of evaluating the "system" that contributed to the failure of a tailings facility is published in the report of the British Columbia chief mine inspector on the Mount Polley tailings failure (*Investigation Report of the Chief Inspector of Mines* 2015). This report, supported by Klohn Crippen Berger (2015), uses a root cause analysis (RCA) approach to identify the specifics and interaction of various factors that played a role in the failure. The outcome was an "RCA event and causal factor tree," which was provided in the report (Figure 1).

The RCA carefully evaluates both technical and organizational factors to consider the structural causes of failure of the perimeter embankment and the breach and release of the water and tailings leading to downstream impacts. Organizational factors are included as far as possible.

The site-specific nature of all tailings failures clearly makes it important to include the local organizational matters in an RCA. This is not a simple matter as there are often more aspects than those exhibited in the records, and that is where the true complexity sets in. It is extremely difficult to know why specific decisions were made over time. For example, there are clear discussions of the technical matters that were documented by the engineers of record; however, there are no records of interactions with the client related to these matters in the files reviewed by the author as part of the panel investigation.

As Dekker (2011) clearly indicates, there are "stories" that are important in evaluating complex systems. The records available in the investigation of immediate (or proximate) technical causes of the failure do not touch on these. Although there is no expectation that this can easily change regarding tailings failures, the importance of diversity in the conformity evaluations implemented by MAC can be a good buffer to reduce the likelihood of future failures.

A FEW COMMENTS ABOUT THE BOEING 737 MAX

Much has been published in the press about the Boeing 737 Max since the crashes in Indonesia and Ethiopia. Many interesting pieces of information have become available in these journalistic investigations as well as congressional testimonies. For the sake of simplicity, the information is provided under the following series of topics.

Boeing Marketing Strategy

The 737 Max was based on the standard 737 platform that was developed in 1968, with a few changes. However, the company strategy was to market the plane as one where pilots did not need new training. This seems to have been an underlying strategy as it could save the company money.

Pursuing this approach, the team at Boeing also worked hard to develop a strategy internally and with the Federal Aviation Administration (FAA) to not include anything in the training manuals about the new instrumentation and flight control systems. The 737 Max went into operation without any information about changes in the pilot manual.

Remaining competitive seemed to have triumphed over the need for absolute safety to the point where internal emails indicated the unwillingness of some Boeing personnel to feel comfortable having their families fly on the plane.

Federal Aviation Administration Changes

The FAA also underwent many regulatory revisions approved by Congress to pass a large part of the responsibility of controls to the manufacturers. This undoubtedly also impacted the overall oversight as well as internal decision-making in the company.

Technical Realities

The new flight control system that was installed to control potential stall conditions, called the *maneuvering characteristics augmentation system*, was based on a software system designed to automatically activate only in the extreme flight situation of a high-speed stall. This extra kick downward of the nose can make the plane feel the same to a pilot as the older model 737s. This software relied on the output of one sensor to be engaged. Although there were two sensors measuring the angle of attack, the output of only one was used by the software instead of the average of the two or some other configuration.

Outcomes

The outcome is that Boeing lost overall market credibility, the share value took a major hit, and the CEO was replaced. The company must figure out how to rebuild credibility and how it can get the 737 Max in the air again and keep manufacturing it after the present stoppage. This represents a major challenge to the company.

Comparisons to Tailings

Although not immediately clear how this relates to tailings facilities, one can come to two comparative outcomes:

1. **The competitive attitude of consulting engineers, especially for new mines.** A strategy exists to "get the foot in the door" and potentially overpromise on the basis of "an old platform" of experience in new mines. This can create an expectation of low-cost tailings facility design and permitting.
2. **The quest for a regulatory environment where the decisions are made by the owner and/or engineer of record without strong oversight and control.** This always creates potential problems.

SOME THOUGHTS

The operations of tailings facilities are successful when the system is carefully managed. This endeavor is a significant challenge, and all efforts must be made to accomplish a truly integrated management system. A full commitment is required, and attention must be given to the large range of guidelines and manuals presently available. At the same time, we need to commit to a better understanding of the system and to integrating systems thinking and diversity into this mix.

REFERENCES

Dekker, S. 2011. *Drift into Failure.* Farnham, UK: Ashgate.

Investigation Report of the Chief Inspector of Mines: Mount Polley Mine Tailings Storage Facility Breach: August 4, 2014. 2015. p. 130, fig. 10.1. https://www2.gov.bc.ca/assets/gov/farming-natural-resources-and-industry/mineral-exploration-mining/documents/directives-alerts-incidents/chief-inspector-s-report-page/m-200_mount_polley_2015-11-30_ci_investigation_report.pdf. Accessed October 2020.

Klohn Crippen Berger. 2015. *Ministry of Energy and Mines: Mount Polley Tailings Dam Failure: A Summary of Opinions in Support of CIM Investigation.* www2.gov.bc.ca/assets/gov/farming-natural-resources-and-industry/mineral-exploration-mining/documents/directives-alerts-incidents/chief-inspector-s-report-page/m-200_mount_polley_2015-08-00_ci_investigation_report_app3.pdf. Accessed October 2020.

MAC (Mining Association of Canada). 2019a. *Developing an Operation, Maintenance, and Surveillance Manual for Tailings and Water Management Facilities*, 2nd ed. Ottawa, ON: Mining Association of Canada.

MAC (Mining Association of Canada). 2019b. *A Guide to the Management of Tailings Facilities.* Ottawa, ON: Mining Association of Canada.

MAC (Mining Association of Canada). 2019c. *Table of Conformance.* https://mining.ca/documents/table-of-conformance-2019/. Accessed June 2020.

Tailings Dam Failures—Acts of Man

Allen H. Gipson Jr.
Gulf Shores, Alabama, United States

INTRODUCTION

The objective of this paper is to look at the lessons learned about the real cause of tailings dam failures. It makes the point that tailings dam failures are the results of acts of man. Initially, a discussion is presented on the past and current thinking related to the cause and classification of tailings dam failures. A brief look at the consequences of tailings dam failures is also presented. This is followed by a new classification system directly relating failures to acts of man. The final section presents takeaways designed to address achieving the goal of zero failures.

FORWARD-LOOKING STATEMENT

The information presented in this paper is predominately supported by publications. In a few cases, I have speculated on additional potential failure causes to demonstrate the main point of this paper. They are believed to be reasonable alternative failure causes. The intent of this paper is not to attribute fault to any specific people but to demonstrate that failures result from acts of man.

BACKGROUND

Tailings dam failures are man-made disasters. In 1994, the U.S. Committee on Large Dams (USCOLD) published a report titled *Tailings Dam Incidents* (USCOLD 1994) that categorized 185 incidents or failures by technical causes including: overtopping, slope instability, earthquakes, foundation, seepage, structural, erosion, mine subsidence, groundwater, and unknown. In 2003, I prepared a paper titled "Tailings Dam Failures: The Human Factor" (Gipson 2003). The paper concluded that the technology exists to achieve the goal of zero incidents or failures, but the human factor inhibits achieving the goal. That paper did not directly relate the failures to specific actions; however, this paper presents a classification that relates the failures directly to human actions.

Papers by others present similar findings. For example, the title question from the 2000 article by Don East asked,

> Will tailings dam failures continue to occur? Probably, as long as the mining industry continues to do business in the same old way. Can they be prevented? The answer is again probably. But it will take a change in the mind-set of mining professionals at all levels in the organization. This is ultimately about management philosophy and the management of risk. Only when mining industry managers at the highest corporate levels accept this fundamental management philosophy and only when they do not just assume, but demand, sound risk-management strategies from all levels of their organization, will the occurrence of tailings dam failures and other environmental incidents be reduced.

And Mike Davies (2002) stated:

> There was a lack of design ability, poor stewardship (construction, operating or closure) or a combination of the two, in each and every case history. If basic design and construction requirements are ignored, a tailings dam's candidacy as a potential failure case is immediate.

FAILURE CONSEQUENCES AND HISTORY

It is educational to briefly look at a few of the consequences of failures to demonstrate the importance of addressing the need to achieve the goal of zero failure. In the United States, 27 coal waste pile failures killed 4,338 people. The most recent, the Buffalo Creek failure, left 118 dead in 1972. Not directly related but similar, the estimated cleanup cost for the 2008 Tennessee Valley Authority (TVA) Kingston Ash Pond failure is US$1.2 billion. We have been living with waste pile and similar ash pond failures for more than 100 years.

The earliest reference I have seen to spontaneous liquefaction is that of Necaxa Dam in Mexico in 1909 (Sherard et al. 1963). Water dam designers stopped using hydraulic fill dams many years ago. The discussions on flow slides caused by spontaneous liquefaction in loose sand foundations in Sherard et al. (1963) describe the causes of the liquefaction as well as the difficulty of predicting the liquefaction susceptibility of soils. More than 100 years later, the mining industry is still plagued with failures resulting from flow liquefaction. Clearly, these events damage the reputation of the mining industry.

ACTS OF MAN—THE REAL CAUSE OF FAILURES

The following are the classifications developed and used in this paper to demonstrate how human actions have led to nearly all failures.

- Incompetence—performing work that is outside one's area of training or expertise or outside one's experience level
- Dereliction of duty—failing to follow instructions or procedures and to review and report monitoring and instrumentation results in a timely fashion
- Fear—performing substandard work or hiding information on critical safety issues for fear of repercussions from the supervisor or client including potential loss of business and the related revenue
- Greed—disregarding prudent engineering, scientific or permitting criteria, or advice to increase profits for personal renumeration
- Ego (state of denial)—using a flawed design concept or analysis and refusing to change rather than openly and transparently considering other variable designs or analyses

Failure Examples Classified by Acts of Man Classification

The following are a few examples of failures with their attributed classifications.

Area C-C'

Area C-C' was a phosphatic clay settling area near Fort Meade, Florida. It covered about half a square mile. The embankment was about 50-ft high with a corrugated metal pipe (CMP) spillway outlet through the base of the dam. It failed suddenly on December 3, 1971, with slimes flowing to the Peace River and into a bay causing huge environmental damage. It failed at the spillway location. Slopes were likely about 2:1 horizontal–vertical. At that time, it was not uncommon practice in the area to design for factors of safety at about 1.2 versus the commonly accepted minimum of 1.5.

By the Acts of Man classification, this falls under incompetence for the likely potential causes of failure, which include designers working outside their area of expertise, slopes too steep, no internal drains to intercept seepage from eroding silty sands used to construct the embankment, questionable compaction, difficult to compact soils beneath the CMP spillway leaving a potential flow path, lack of regulatory requirements, and likely insufficient construction monitoring or inspection protocol.

For design of the next settling area, which had a maximum height of 75 ft, an experienced and qualified designer was retained. Also, the state had developed new regulations that included technical requirements and operator and inspector training as well as frequent dam inspections. The new design included 3:1 or 5:1 exterior slopes, internal chimney and blanket drains, a minimum 5 ft of freeboard, a maximum 5-ft water pond depth, CMP decant outlet pipe bedded in concrete, and continuous inspection by the design engineer during construction. An operator training and inspection program was initiated. The dam performed well.

Dam No. 3

Dam No. 3 failed on October 18, 1980. It was a 220-ft-high cyclone sand dam. The consultant hired by the insurance company seeking damages attributed the failure to "stability failure through the dam due to rapid increase in tailings height causing high internal pore pressures in tailings from rapid filling…" In USCOLD (1994), the cause is described as "sliding of embankment slope." Another potential cause considered was not performing a routine inspection of the tailings delivery line located on the downstream slope just below the crest as the failure occurred early in the morning on a stormy night. Also, there was an indication of a small earthquake or blast before the failure.

No matter how it is viewed, it is either incompetence or dereliction of duty. It could have been the incompetence of the designer to specify a maximum rate of rise, incompetence or dereliction of duty of the operator to limit the rate of rise, incompetence of the designer to consider seismic liquefaction, or dereliction of duty of the operator to inspect the tailings delivery header.

Gypsum Stack

The foundation of this gypsum stack failed because of a flow failure following wetting of collapsible soils after initiation of process water storage for start-up. The preliminary geotechnical investigation identified collapsible soils beneath the embankment and recommended they be removed and replaced with compacted fill. Collapsible soils are common in the midwestern United States.

The owner switched designers after the preliminary design. The new designer was from the southeast where collapsible soils are not found. The preliminary recommendations for removal and replacement of the collapsible soils were not followed. The soils underwent flow liquefaction on the first filling with water for start-up.

The potential Acts of Man classifications include incompetence of the designer by taking an assignment outside the designer's area of expertise and experience and ignoring the preliminary recommendations. Greed stemming from potential cost savings by not removing and replacing the collapsible soils is also a consideration.

Keystone Mine

Four tailings dams were built at the Keystone mine between 1955 and 1975. They were built in an echelon series on a fairly steep slope above the pristine fishery Coal Creek. The dams had a maximum height of 40 ft. The slopes were about 1.5:1. Deposition was from upstream to the downstream crest, placing the weaker slimes and water recovery system at the dam crest. Dams 2 and 4 failed in 1975. The tailings from Dam 4 flowed into Coal Creek instigating a lawsuit by the State of Colorado (Gipson 1993).

The potential Acts of Man leading to the failure include incompetence on the part of the designer not having the necessary qualifications and experience to design tailings dams, and/or greed on the owner's part by hiring an inexperienced firm to save money, and dereliction of duty and/or incompetence on the part of the regulators for not providing adequate review and inspection.

South Africa

This tailings dam in South Africa was overtopped as a result of unauthorized deposition, resulting in the death of 17 people, and dereliction of duty was the cause of failure (Wagener 1997):

> The dam had a low freeboard due to unauthorized slime being deposited and water stored on compartment 4A in spite of the decision to discontinue sliming on the dam. The trigger, which finally caused the disaster, was a thunderstorm of only 50 mm. The dam overtopped following the rain and this caused erosion and slip failures which moved up the slope in domino fashion eventually reaching the top of the dam and leading to the massive slope failure and mud-rush engulfing the town.

Other Potential Problems

Following is a list of some other things that can lead to failures in tailings dams:

- Changing designers without the new designer fully understanding the existing designs
- Accepting a low bid without regard to qualifications and experience
- Making design changes without review and approval of the designer

- Cutting corners, for example, changing the design to reduce costs (if the project is not viable without violating proven technology, then it is not a viable project)
- Lack of regulatory requirements and oversight
- Offering "generous bonuses for mine managers, rewarding them for reducing costs and increasing production" (Lavoie 2019)

PROFESSIONAL RESPONSIBILITY

Based on my experience, establishing a culture based on professional responsibility goes a long way to meeting the ultimate goal of zero harm to people and the environment. But what is professional responsibility? Professor of Rock Mechanics at the Colorado School of Mines John F. Able Jr. put it to me in the most basic of terms: "I haven't killed anyone yet, and I don't intend to start now." And while developing design criteria for dams related to a mining project in Peru, we came up with the following: "When constructing a mining facility, it should be done in a manner that does not increase the risk to those in the area or to the environment."

Guidelines from the Canadian Dam Association (CDA) state, "The public and the environment shall be protected from the effects of dam failure, as well as release of any or all fluids behind a dam, such that the risks are kept as low as reasonably practicable" (CDA 2007). This is a start, but "as low as reasonably practicable" leaves a lot of room for interpretation. The Global Tailings Review statement is much more definitive (GTR 2019):

> Ultimately, operators are required to strive towards zero harm to people and the environment—with zero tolerance for human fatality. The mining industry is encouraged to invest in new technologies and safer mining methods to achieve this goal.

The professional engineer's obligation to society as stated in the Colorado Revised Statutes (2019) states

> ...In order to safeguard life, health and property and to promote the public welfare, the practice of engineering is declared to be subject to regulation in public interest. It shall be deemed that the right to engage in the practice of engineering is a privilege...that the profession involves personal skill and presupposes a period of intensive preparation, internship, due examination, and admission...

TAKEAWAYS

Tailings dam failures are the direct result of acts of man. The technology exists to design, operate, and close tailings dams to achieve the goal of zero failures. *We are really dealing with lessons relearned, not lessons learned.*

What Can We Do to Eliminate Failures?

The *Global Tailings Standard* (GTR 2019, 2020) and the *Proposed Best Practices for the Engineer of Record (EoR) for Tailings Dams* (GBA 2018) are a good first step in learning about and implementing best practices. Additional recommendations include the following:

- Draft a universal design, construction, operation, inspection and monitoring, and closure manual. Develop a comprehensive universal design manual for international use, which can provide design information on tailings dams from cradle to grave. Examples of relative failures should be included to demonstrate lessons we have already learned to prevent them from becoming lessons relearned.
- Develop a tailings dam failure database as a reference source and training tool for practitioners similar to the inventory of tailings dam incidents compiled by USCOLD (1994), including a description of the technical cause of failure. By studying failures, we can learn not to make the same mistakes again and again.
- Create training and certification programs for the engineer of record, designer of record, operator tailings engineer, operator maintenance and inspection personnel, and regulators. Formal programs go a long way to providing the necessary training and developing a knowledgeable workforce of qualified, experienced, and certified professionals.
- Consider more stringent design requirements for tailings dams. Tailings dams involve a much greater degree of complexity than water storage dams. The consequences of tailings dam failures are inherently greater than water dams. In emergency situations, water can be released from water storage dams through the outlet works to reduce the head on the affected area. Not so with tailings dams. When tailings dams fail, they release tailings that need to be removed or covered in place. Also, in some cases, the fluids and material released are toxic. At closure, tailings dams cannot simply be breached and contoured to blend into the surrounding landscape. Tailings dams will be there in perpetuity and need to be maintained in perpetuity.

CONCLUSION

Simply put, owners, operators, engineers, scientists, and regulators need to be professionally responsible for the safe design, construction, operation, closure, and long-term maintenance of tailings dams. Personnel need to be trustworthy and have the necessary education, training, experience, certification, honesty, integrity, and sound moral and ethical values to achieve the goal of zero harm to the environment—with zero tolerance for human fatality. Governance programs and introduction of the engineer of record are commendable, but it boils down to the individuals involved and the culture of the organizations to achieve the goal of zero failures.

REFERENCES

CDA (Canadian Dam Association). 2007. *Dam Safety Guidelines*. Toronto, ON: Canadian Dam Association.

Colorado Revised Statutes. 2019. Articles 120 and 121: Engineers, Surveyors, and Architects. Title 12: Professions and Occupations. Denver.

Davies, M.P. 2002. Tailings impoundment failures: Are geotechnical engineers listening? *Waste Geotechnics, Geotechnical News* (September). https://miningquiz.com/pdf/Impoundments/Dam_failuresDavies2002.pdf. Accessed June 2020.

East, D.R. 2000. Tailings dam failures—Why do they continue to occur? *Mining Engineering* 52(12):57–60.

GBA (Geoprofessional Business Association). 2018. *Proposed Best Practices for the Engineer of Record (EoR) for Tailings Dams*. Rockville, MD: GBA.

Gipson, A.H. Jr. 1993. Keystone mine tailings dam reclamation. In *Air Pollution Control: Proceedings: Papers from the 86th Annual Meeting and Exhibition of the Air & Waste Management Association*. Pittsburgh: Air & Waste Management Association.

Gipson, A.H. Jr. 2003. Tailings dam failures: The human factor. In *Tailings and Mine Waste '03*. Rotterdam, The Netherlands: A.A. Balkema.

GTR (Global Tailings Review). 2019. *Global Tailings Standard: Draft for Public Consultation*. https://globaltailingsreview.org/wp-content/uploads/2019/11/EN-Global-Tailings-Standard_CONSULTATION-DRAFT.pdf. Accessed November 2020.

GTR (Global Tailings Review). 2020. *Global Industry Standard on Tailings Management*. https://globaltailingsreview.org/wp-content/uploads/2020/08/global-industry-standard_EN.pdf. Accessed November 2020.

Lavoie, J. 2019. Tailings dam failures linked to hefty bonuses for mine managers: Report. *The Narwhal* (June 24). https://thenarwhal.ca/tailings-dam-failures-linked-to-hefty-bonuses-for-mine-managers-report/. Accessed June 2020.

Sherard, J.L., Woodward, R.J., Gizienski, S.F., et al. 1963. *Earth and Earth Rock Dams: Engineering Problems of Design and Construction*. New York: Wiley.

USCOLD (U.S. Committee on Large Dams). 1994. *Tailings Dam Incidents*. Denver: U.S. Committee on Large Dams.

Wagener, F. 1997. The Merriespruit slimes dam failure: Overview and lessons learnt. *SAICE Journal* 39(3):11–15.

Mount Polley Mine Tailings Spill: Five Years Later

C.D. (Lyn) Anglin
Anglin & Associates, North Vancouver, British Columbia, Canada

INTRODUCTION

Five years after a glaciolacustrine layer in the natural materials under the foundation of the tailings dam at the Mount Polley mine (MPM) failed, and 17 million m^3 (cubic meters) of water and 8 million m^3 of tailings escaped the tailings storage facility (TSF), significant progress has been made in remediating the affected environment. The company has currently invested about Can$70 million on repair and cleanup, including environmental impact assessments, monitoring studies, and remediation of impacted areas. The Hazeltine Creek channel was completely rebuilt by May 2015 to control erosion. More than 600,000 native plants, shrubs, and trees have been planted in the creek corridor. In upper to middle Hazeltine Creek, more than 6 km of new trout spawning and rearing habitat has been installed. In lower Edney Creek, salmon spawning habitat has been restored and successfully used by salmon spawners. The remediation work is ongoing, with lower Hazeltine Creek as the focus of work in 2020. Geochemical studies, sediment and water monitoring, and benthic species monitoring show that the tailings that were deposited at the bottom of Quesnel Lake are physically and chemically stable and are not releasing metals to the lake water. Monitoring indicates that the terrestrial and aquatic ecosystems affected by the spill are showing positive signs of recovery. Communications with stakeholders and First Nations have been maintained through regular community meetings, community newsletters, committee meetings, and site tours with regulators and indigenous and community representatives.

This paper presents a brief overview of the MPM's emergency response actions, a summary of the impact and risk assessments completed, the company's repair and remediation efforts, the communications and stakeholder engagement that the company undertook, and the status of the remediation as of the end of 2019.

BACKGROUND ON THE MOUNT POLLEY MINE

Mount Polley Mining Corporation (MPMC), a wholly owned subsidiary of Imperial Metals Corporation (IMC), owns and operates the Mount Polley copper-gold mine located in the Cariboo Regional District, 56 km northeast of Williams Lake, in the interior rainforest of central British Columbia (Figure 1).

OVERVIEW OF THE EVENT

Early on August 4, 2014, at 1:10 AM Pacific Time, a breach of the MPM TSF perimeter embankment occurred, which resulted in the release of approximately 25 million m^3 of tailings solids, supernatant and interstitial pore

Courtesy of Imperial Metals

Figure 1 Mount Polley mine location in central British Columbia, Canada

water, and construction (dam embankment) materials (Table 1; Figure 2). Construction materials, water, and tailings spilled into the adjacent Polley Lake, then down the Hazeltine Creek corridor, to the mouth of Edney Creek, and into Quesnel Lake about 8 km away. Hazeltine and lower Edney creeks were scoured in the process, causing eroded materials, tailings, and woody debris to be deposited in Polley and Quesnel lakes.

Table 1 Estimated amounts of displaced materials

Material	Amount, million m^3
Supernatant water	10.6
Tailings slurry	
Tailings solids	7.3
Interstitial water	6.5
Construction materials	0.6
Total materials	25.0

Courtesy of Imperial Metals

Figure 2 Mount Polley Mining Corporation tailings storage facility perimeter embankment of area of failure on August 4, 2014, looking south

Cause of Failure Assessment

Independent experts, and the provincial government, examined the August 2014 MPM embankment breach. A three-member expert panel was established by the government of British Columbia through the Ministry of Energy and Mines (MEM), together with the Williams Lake Indian Band (WLIB) and the Soda Creek Indian Band (SCIB) (Morgenstern et al. 2015). An independent geotechnical engineer was retained by MPMC to investigate the cause of the tailings dam failure. Detailed site investigation, laboratory testing, and review of historical site characterization data resulted in a consistent finding that the breach of the MPM TSF arose from a failure in a glaciolacustrine unit located 8 to 10 m (meters) below the base of the embankment in native glacial sediments.

EMERGENCY RESPONSE

The MPMC emergency response plan was enabled with the assistance of off-site government and community authorities. MPM immediately ceased operations and initiated a series of actions to ensure human health and safety and mitigate further environmental impacts. Mine personnel contacted Emergency Management British Columbia (EMBC), MEM, Royal Canadian Mounted Police, First Nations, senior mine staff, and corporate officers. The area was secured, and the local residents were notified. There was no loss of human life associated with the spill event or with any of the response activities. The British Columbia Ministry of Environment (ENV) issued a Pollution Abatement Order (PAO) under the British Columbia Environmental Management Act on August 5. The PAO directed MPMC to implement measures and submit documentation describing its response and to communicate to the ENV regarding response progress.

Impact Area

The breach resulted in the release of tailings, supernatant and interstitial water, and construction material from the TSF into the environment, as well as significant physical damage to the immediate environment. Tailings were deposited in the area immediately adjacent to the TSF and in Polley Lake, and a large amount of tailings and native material that was scoured out of the Hazeltine Creek valley, and from the mouth of Edney Creek, were transported into and deposited on the bottom of Quesnel Lake. This area has a long history of resource extraction activity. It is the site of extensive historic and ongoing placer and hard rock mining, forestry, and agriculture. It is also an important area for recreational fishing, seasonal and year-round homes, and hosts commercial fishing, hunting, and guiding lodges (Figure 3).

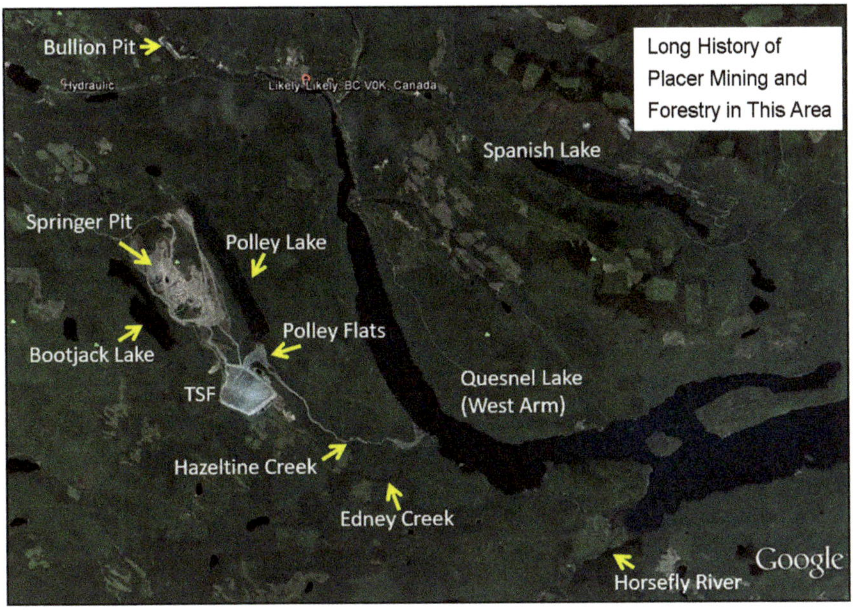

Courtesy of GoogleEarth

Figure 3 Annotated image showing the regional area of impact around Mount Polley mine in central British Columbia, Canada

The total terrestrial area impacted by the spill is estimated to be approximately 2.4 km^2, with 1.36 km^2 of riparian and forested area scoured of vegetation and topsoil (floodplain zone), and a further 1.0 km^2 area where tailings were deposited on top of relatively undisturbed forest floor. Approximately 12.8 million m^3 of tailings are estimated to have been discharged to Quesnel Lake (plus an additional 5.8 million m^3 of native soil and TSF water), with the bulk of this material estimated to have settled in the West Basin of Quesnel Lake over an estimated area of 1.8 km^2. An additional 1.6 million m^3 of tailings were deposited in the Polley Flats area and Hazeltine Creek corridor (Golder 2019).

Mount Polley Mine Immediate Response

Resources were mobilized to identify and/or minimize risk:

- Implement safe work plans for the impact zone.
- Sample water and contain woody debris floating in Quesnel Lake.
- Supply bottled drinking water to those drawing their drinking water directly from Quesnel Lake.
- Assess the stability of the Polley Lake outflow.

- Stabilize the sediment plug at Polley Lake outflow and pump out 6 million m^3 of excess water.
- Contain the remaining tailings in the TSF.
- Capture and contain mine contact water to the mine site.
- Control erosion potential and sedimentation.
- Conduct a post-event environmental impact assessment.

A large amount of vegetation (trees, shrubs, etc.) was scoured out of the creek valley during the spill event and deposited into Quesnel Lake, creating a significant amount of floating debris on the lake. Work began immediately to contain woody debris to prevent damage to the wooden bridge crossing the Quesnel River at Likely (at the outlet of Quesnel Lake) and reduce boating hazards. Floating debris was corralled in booms then tugged to a reload for dewatering, sales, chipping, or burning (Figure 4). Beaches were cleaned, starting with those located near private residences.

A substantial amount of water from the TSF was transported into Polley Lake, and a dam of sediment and other debris at the lake outlet artificially raised the level of Polley Lake. A rock fill buttress was built across the Polley

Courtesy of Imperial Metals

Figure 4 Wood debris being collected on Quesnel Lake, August 2014

Lake outlet, and the water level was pumped down 1.7 m to the pre-breach lake level. Lowering the level of Polley Lake was deemed to be critical to ensure the safety of personnel working below the "Polley Plug" in the creek valley who were investigating the impact of the spill, and those working at the base of the TSF who were investigating the cause of the failure. A water control structure was later installed at the lake outlet (Polley Lake weir) to control flow rate into Hazeltine Creek to prevent erosion and allow creek reconstruction.

A rock fill dike was built inside the TSF across the breach to contain the approximately 80% of tailings solids remaining in the TSF and to allow for safe access below the breach for remediation work. The rock fill dike was completed on September 5. Prior to the breach, the site water system reported to the TSF, so a breach and central collection sump was constructed to capture and pump water through relocated pipelines to Springer pit. Contact water was fully contained on September 9. Sediment control ponds were constructed in lower Hazeltine Creek and put into operation in December 2014 to capture suspended solids before water flowing down the creek entered Quesnel Lake. Two forest service road bridges that crossed Hazeltine Creek had been damaged in the debris flow. These bridges were replaced with temporary structures in fall 2014, with permanent replacements installed in 2016.

Despite the drinking water and fish consumption advisories being lifted by the British Columbia Interior Health Authority (and supported by the First Nations Health Authority), MPMC organized testing of home water quality and installation of water filtration systems, including supplying water filters to residents who drew water from Quesnel Lake. By September 25, more than 50% of the homes in the Likely area had filters installed by MPMC contractors. In response to a request from a First Nations band representative, MPMC also provided cans of salmon for band members who were concerned about potential impacts on salmon downstream.

IMPACT ASSESSMENT AND REHABILITATION AND REMEDIATION STRATEGY

MPMC, working with support from their consulting team, developed a rehabilitation and remediation strategy (RRS) to guide the post-event environmental monitoring and remediation of terrestrial and aquatic systems (McMahen and Hughes 2016). For a comprehensive overview of the response and approach to remediation, see "Mount Polley Mine August 2014 Tailings Dam Embankment Breach: Response and Approach to Remediation" by McMahen and Hughes (2016). Much of the following section is summarized from this paper. The areas of impact were delineated into subareas for the purposes of impact and risk assessments and remediation and monitoring planning (Figure 5).

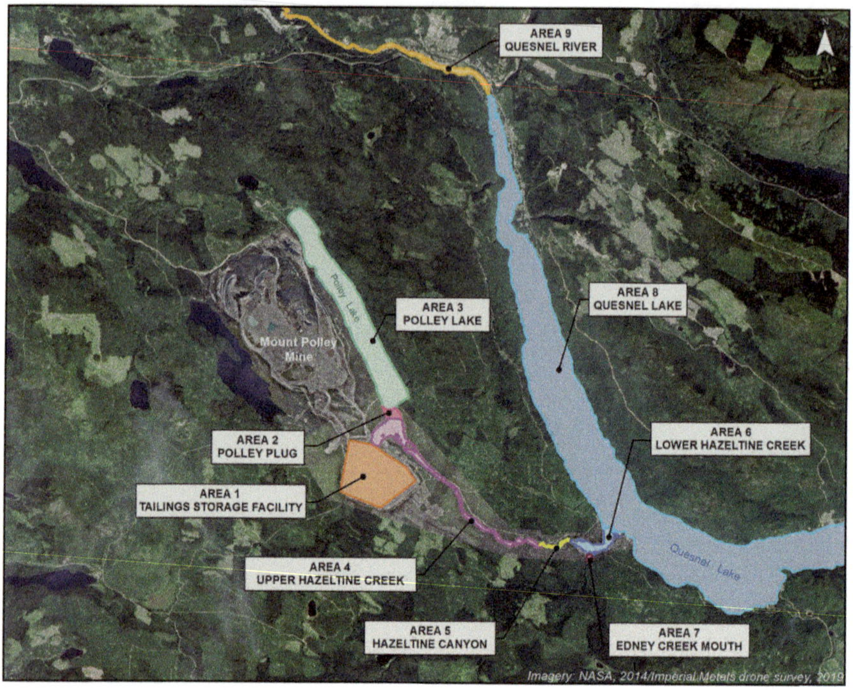

Courtesy of Imperial Metals

Figure 5 Mount Polley mine tailings dam breach—risk assessment and remediation areas

Phase 1: Initial Response

Phase 1 focused on addressing health and safety concerns, including establishing safe working conditions and restricting public access to impacted areas, stopping inputs of material to the surrounding environment, and urgent restoration work primarily to control erosion and stabilize the impacted sites. (Much of this phase of the RRS is discussed in the preceding section "Emergency Response.")

A critical part of the response involved controlling the erosion and transport of deposited dam outwash materials and material eroding from over-steepened scoured creek banks from entering Quesnel Lake. An extensive program of source control measures was implemented over the following weeks and months, including (McMahen and Hughes 2016)

- Modifications and upgrades to the mine water collection and conveyance systems to direct water away from the TSF and into the Springer pit;
- Construction of the temporary rock fill dike to contain remaining tailings in the TSF;

- Repair of the perimeter embankment breach to allow use of the TSF for management of spring freshet flows and completed May 2015;
- Helicopter grass seeding along Hazeltine Creek and on exposed tailings;
- Construction of sediment control ponds at the bottom of Hazeltine Creek (following extensive archaeological studies) to reduce potential suspended solids inputs to Quesnel Lake during stream stabilization work; and
- Initiation of a field-engineered stream reconstruction and rehabilitation program involving removal of tailings, excavation to competent ground, and construction of a rock-armored channels for both Hazeltine and Edney creeks (Bronsro et al. 2016).

A large-scale environmental monitoring program was designed and formalized in the weeks following the breach that became the focus of phase 2.

Phase 2: Short-Term Impact Assessment

Phase 2 included reporting on immediate impacts, that is, the physical, chemical, and biological effects in the first six to eight months post-breach; identifying data gaps and developing sampling plans to address these gaps; and developing a conceptual site model for understanding and investigating the extent of the impacts and associated risks to ecosystem components. This phase of the RRS program included assessment of water quality, sediment quality, soil quality, geochemistry, hydrology, geomorphology, limnology, fish and fish habitat, aquatic toxicology, vegetation, and terrestrial wildlife. The results of these studies are presented in the post-event environmental impact assessment report (PEEIAR) (Miller et al. 2015).

Phase 3: Detailed Site Investigation, Risk Assessment, and Remediation Options

Phase 3 followed a contaminated site regulations–based investigation approach and included three subphases, which built on the results of the phase 2 PEEIAR:

1. Detailed site investigation includes characterization of soil quality, groundwater quality, surface water quality, sediment quality, and site geochemistry.
2. Human health and ecological problem formulation include investigations and technical reports on plant and invertebrate tissues, plant toxicity testing, zooplankton community and tissues, and fish tissues.
3. Risk assessment and performance verification plan involve risk assessments for human, aquatic, and terrestrial receptors, and preparation of the remediation plan.

Phase 4: Implementation of Remedial Actions and Monitoring

Phase 4 activities, although listed as the final stage of the RRS, were undertaken throughout the response process, with remedial actions and monitoring beginning the day of the event. As progress was made in all phases of the response, the results of studies and monitoring undertaken as part of the RRS were used to inform the remedial actions and further monitoring activity. This iterative approach combined and built on existing response processes to balance the demands of urgent actions, such as source control, with the methodical assessments needed to meet the standards required for regulatory approvals. Importantly, MPMC worked closely with the ENV Statutory Decision Maker throughout this process to ensure that regulatory expectations were being met, and necessary modifications to the response plan were made on a timely basis.

The first PEEIAR was submitted to ENV in June 2015 (Miller et al. 2015) and provided the results of the initial studies conducted to evaluate the environmental impacts of the breach. An updated PEEIAR was requested by ENV and completed in June 2016 (Miller and Zapf-Gilje 2016). Subsequent studies and reports, including the *Human Health Risk Assessment* (Golder 2017b), the *Ecological Risk Assessment* (ERA) (Golder 2017a), and the *Final Remediation Plan* submitted in March 2019 (Golder 2019), built on these impact assessment studies and many other technical reports, memos, monitoring results, and analyses.

The *Human Health Risk Assessment* (Golder 2017b) included the assessment of risks to subsistence and traditional land users, hikers and campers, ATV and snowmobile users, boaters and kayakers, sport fishers, hunters and trappers, Quesnel Lake residents, and loggers and miners and rehabilitation workers. Exposures to Al, Cu, and V were evaluated, and calculated hazard indexes were determined to be below the threshold of 1 for each of the exposure pathways assessed. Human health risks associated with the TSF embankment breach are considered to be low and not a risk to human health.

The ERA (Golder 2017a) determined that copper was the primary constituent of potential concern associated with tailings, and no other chemical constituents were found to be a significant ecological risk. The risks associated with copper are considered to be low. Tailings have structural (grain size and homogeneity) and nutrient (organic carbon) deficiencies that are important non-contaminant stressors, but overall ecological risks are considered low to moderate. Where risks are considered moderate, they are associated primarily with disruption to physical substrate, both scouring and deposition. (Mount Polley's mine site water is not considered toxic to sensitive aquatic life.) ERA recommendations for ongoing monitoring of vegetation in the Hazeltine Creek corridor and aquatic life in Hazeltine Creek, Polley Lake, and Quesnel

Lake were incorporated into the comprehensive environmental monitoring plan (CEMP).

Remediation Committees and Working Groups

MPMC, working together with regulators, government agencies, First Nations partners, staff, and consulting firms, developed and enacted plans in keeping with the regulatory orders outlined earlier. As part of working on these plans, and reviewing results, several technical committees and working groups were created. In particular, the Habitat Remediation Working Group (HRWG), which was initiated by the Department of Fisheries and Oceans (DFO) and hosted by MPMC, includes First Nations, British Columbia Forest Land and Natural Resources Operations (FLNRO), DFO, ENV, and company consultant representatives and was originally focused on fish habitat remediation. This group has been effective and an invaluable contributor to the planning and design of the overall habitat remediation work done by MPMC. The HRWG is continuing to meet and works to oversee the terrestrial remediation and fish habitat reconstruction work.

Independent Research

As part of the impact assessments and supporting a broad understanding of the effects of the spill, MPMC directly and indirectly supported academic research groups carrying out independent studies on various aspects of the spill. MPMC also supported a Quesnel Lake monitoring buoys program. With ENV, the mine jointly purchased scientific research instruments to be installed on five buoyed moorings in Quesnel Lake by DFO researchers, with assistance from the University of Northern British Columbia Quesnel River Research Centre and the University of British Columbia staff members. Starting in 2013, MPMC also provided financial support to FLNRO for a resident fish telemetry tagging program to monitor lake and rainbow trout, burbot, and kokanee in Quesnel Lake. MPMC continued to support this program for several years post-TSF spill. Independent studies have also been carried out by a number of academic researchers with in-kind support from MPMC, and several papers have been published on their results, including Garris et al. (2018), Byrne et al. (2018), Cuervo et al. (2017), and Hudson-Edwards et al. (2019).

PUBLIC ENGAGEMENT AND COMMUNICATION

After the breach, MPMC immediately engaged with local communities in the area. About 40 formal community meetings were organized and hosted by MPMC; more than half of them were held in Likely, the community in closest proximity to MPM. Several meetings were held in Likely in the first week after the breach, then they were held weekly, then bi-weekly. In 2015, the community meetings were held monthly and then quarterly. Meetings were also held in the communities of Horsefly, Big Lake, Quesnel, and Williams Lake. These meetings provided an opportunity for MPMC to update local residents on progress at the site on cleanup, repair, and remediation work, and the results of research and monitoring programs. They also provided residents with the opportunity to ask questions about operations, spill impact, and risk assessment work, and the remediation planning and implementation. Guest speakers often included consultants and representatives from provincial and federal government agencies. In addition, several MPMC and IMC employees and consultants have given presentations at technical meetings, including at the Tailings and Mine Waste Conference in Keystone, Colorado, in 2016 (extended abstracts are included in the conference proceedings volume).

The mine also regularly held meetings with First Nations in their respective communities to present on the progress of the remediation work, the research on the impacts of the spill, human health and environmental risk assessments, and remediation plan development. These meetings provided an opportunity for members of the First Nations communities to directly ask questions of the mine staff and the consultants working on the response and remediation. These meetings were convened in addition to the regular meetings of the Implementation Committee that were held to discuss the mine's participation agreements (PAs) with First Nations. A PA was signed between MPMC and the WLIB in 2011 and renewed in 2016, and between MPMC and the SCIB in 2012 and renewed in 2017. MPMC also continues to hold quarterly meetings of its public liaison committee (PLC), which had originally been established in 1996. Staff also prepared regular community updates on activities at the mine, particularly progress on repair and remediation work. These updates were distributed in hard copy through the local post office and were made available online. At time of writing, these updates are still available on the Mount Polley pages of the IMC website.

Mount Polley staff have provided numerous site tours of the mine and areas impacted by the 2014 breach. Participants have included local residents, First Nations, school groups, government and community group representatives, other mining companies, professional associations, and media. In 2018, MPMC hosted the Technical and Research Committee on Reclamation Symposium held in Williams Lake and toured almost 230 conference participants around the site. Most recently, in August 2019, company staff led tours for three members of the Global Tailings Review (GTR) team, including Bruno Oberle, chair of the GTR, along with elected and community representatives from the area. In September 2019, mine staff toured a group

Courtesy of Imperial Metals

Figure 6 Teachers' tour of the Mount Polley remediation, September 2019

of local high school and elementary school teachers around the site to view the remediation. Figure 6 shows the MineralsEd Williams Lake teachers observing the fish ladder in upper Hazeltine Creek from the Polley Lake weir. (MineralsEd is a non-profit group that provides educational opportunities to students and teachers in British Columbia about rocks, minerals, and the mineral sector.)

COMPLETED REMEDIATION WORK

MPMC has spent approximately Can$70 million on cleanup, environmental impact, and risk assessment studies, and remediation and monitoring of areas impacted by the spill. The cleanup work completed to date includes

- Collecting woody debris from Quesnel Lake and Hazeltine Creek valley;
- Cleaning up woody debris from Quesnel Lake shoreline, particularly in vicinity of residents' homes;
- Removing tailings from along Hazeltine and lower Edney creeks;

- Rebuilding Hazeltine Creek and the mouth of Edney Creek, including re-sloping of over-steepened banks and ground preparation for planting (Bronsro et al. 2016);
- Repairing the impacted Quesnel Lake shoreline at the mouth of Edney (Hazeltine) Creek;
- Repairing the outflow of Polley Lake, including installation of a water control structure with fish ladder, and cleanup along Polley Lake shoreline;
- Annually gathering and spreading seeds;
- Collecting willow cuttings on-site for use in remediation as whips, stakes, and wattles;
- Planting more than 600,000 native shrubs and trees in riparian and uplands areas;
- Installing approximately 6 km of new fish spawning and rearing habitat in Hazeltine Creek and Edney Creek; and
- Building a rainbow trout (RBT) hatchery on-site in 2018 to raise RBT for Polley Lake.

MPMC has removed as much of the spilled tailings material as was practical (i.e., would not cause additional environmental damage by its removal) and returned most of this material to the TSF or used it in construction of new buttresses on the TSF. A small amount of this material has been stockpiled for potential use in remediation. Previous studies of MPM's tailings have shown them to be unreactive in the environment, and thus suitable as construction material and for use in reclamation work.

The mouth and lower 800 m of Edney Creek were damaged during the spill event. Because this creek directly links to Quesnel Lake (Hazeltine Creek flowed into Edney before reaching Quesnel Lake), it was prioritized for repair in winter 2014 so that spawning habitat in this watershed would be available for any spawners returning in 2015. Fish have been observed using Edney Creek since the summer of 2015. On one day in September 2019, observers reported approximately 30 sockeye salmon migrating through and spawning in the rebuilt part of the creek.

In addition, the mine has installed new fish habitat, specifically a spawning and rearing habitat for RBT, in upper to middle Hazeltine Creek. The upper 5 km of Hazeltine Creek fish habitat construction has been completed. Fish from Polley Lake, most importantly the RBT, were allowed back into the rebuilt Hazeltine Creek in May 2018, where they successfully spawned. Thousands of RBT fry returned to Polley Lake. Spawning and rearing of RBT in Hazeltine Creek in 2019 also appears to have been successful. To augment

Courtesy of Imperial Metals

Figure 7 Progressive remediation work in lower Hazeltine Creek in (A) 2014, (B) 2015, and (C) 2016

recovery of the RBT population in Polley Lake, MPMC also built and operated a hatchery at the mine in 2018. More than 11,000 hatchery-raised RBT were released into Polley Lake in September 2018, with the assistance of students and teachers from a local school in Williams Lake. It is estimated that approximately 5,000 RBT returned to Hazeltine Creek to spawn naturally in 2018, and fall survival surveys included counts in excess of 50,000 fry that returned to Polley Lake from the creek.

In lower Hazeltine Creek (Figure 7; for more photographs, see IMC 2020), the remediation work is not yet complete, so fish are still excluded from this part of the creek. The early repair of lower Edney Creek meant that any spawners returning to the watershed in fall 2015 had access to spawning grounds. In addition, salmon spawning was not significant in Hazeltine Creek prior to the spill as there is a rock canyon and waterfall that forms a natural barrier to their upstream migration about 800 m from the lakeshore. In the years prior to the spill, most of lower Hazeltine was also inaccessible to salmon because of a beaver dam built across the creek.

NEXT STEPS

Remediation and monitoring, as well as community engagement, continue to be priorities.

Ongoing Remediation and Monitoring

MPMC's remediation plan was accepted by ENV, and the PAO was deemed no longer in effect in September 2019. Remediation work on lower Hazeltine Creek, fish habitat reconstruction, and creek valley revegetation work is continuing. Designs for lower Hazeltine Creek and the removal of the sedimentation ponds, or their incorporation into the creek remediation, are presently underway and will be presented to the HRWG for review and approval. Assessment of habitat productivity and discussions of offsetting projects are underway. The mine also continues to follow their CEMP of impacted areas.

Ongoing Community Engagement

Community engagement work also continues with quarterly meetings of the PLC and an annual community site tour. The mine accommodates requests for tours from professional, academic, school, or other groups, when possible. The Implementation Committee also continues to meet when necessary to discuss aspects of the PAs.

ACKNOWLEDGMENTS

Input to this paper and background information has been provided directly or indirectly from past conversations with Don Parsons, Chief Operating

Officer, Imperial Metals; Colleen Hughes, Mount Polley Mining Corporation (MPMC) Environmental Supervisor (now with DeBeers); Katie McMahen, MPMC Environmental Team (now a Ph.D. student at the University of British Columbia); Luke Moger, MPMC Project Manager (now a consultant with Weymark Consulting Ltd.); Shauna Litke, MPMC Environmental Team (now a consultant with SRK); Art Frye, MPMC Operations Manager (now Imperial Metals Mine Planning Team); Gabriel Holmes, MPMC Environmental Department; and a team of consultants, including Golder Associates, led by Lee Nikl; SRK consultants Chris Kennedy (now at Agnico Eagle) and Steve Day; Minnow consultants, led by Pierre Stecko; and many additional experts from TetraTech, SNC Lavalin, EnviroWest, DWB, and others. Their efforts, and those of all the staff and employees who worked at Mount Polley, are gratefully acknowledged. Particular thanks go to Art Frye, Sabine Goetz, IMC Investor Relations, and Melissa Darnley, GIS and cartography consultant, for their assistance in procuring photos and images for this abstract, and to Brian Kynoch, President of IMC for permission to use these images and other information provided by MPMC and IMC.

DISCLAIMER

As former Chief Scientific Officer of IMC (September 2014 to December 2018), much of this paper includes direct observation of the company's efforts. Additional information has been kindly provided by former colleagues at IMC and the MPMC. I have tried to include all of those who have contributed information to this article in the acknowledgments; however, I am solely responsible for any opinions, errors, or omissions in this paper.

REFERENCES

Bronsro, A., Ogilve, J., Adams, M., et al. 2016. River rehabilitation following a dam breach. In *The Technical and Research Committee on Reclamation, Symposium Proceedings*. Victoria, BC: Ministry of Energy, Mines and Petroleum Resources.

Byrne, P., Hudson-Edwards, K.A., Bird, G., et al. 2018. Water quality impacts and river system recovery following the 2014 Mount Polley mine tailings dam spill, British Columbia, Canada. *Applied Geochemistry* 91:64–74.

Cuervo, V., Burge, L., Beaugrand, H., et al. 2017. Downstream geomorphic response of the 2014 Mount Polley tailings dam failure, British Columbia. In *Advancing Culture of Living with Landslides: 4th World Landslide Forum*. Edited by M. Mikoš, Vít Vilímek, Yueping Yin, et al. Cham, Switzerland: Springer International.

Garris, H.W., Baldwin, S.A., Taylor, J., et al. 2018. Short-term microbial effects of a large-scale mine-tailings storage facility collapse on the local natural environment. *PLoS One* 13(4):e0196032.

Golder. 2017a. *Mount Polley Rehabilitation and Remediation Strategy: Ecological Risk Assessment*. Vancouver, BC: Golder Associates Ltd.

Golder. 2017b. *Mount Polley Rehabilitation and Remediation Strategy: Human Health Risk Assessment.* Vancouver, BC: Golder Associates Ltd.

Golder. 2019. *Remediation Plan: Mount Polley Mine Perimeter Embankment Breach.* Vancouver, BC: Golder Associates Ltd.

Hudson-Edwards, K.A., Byrne, P., Bird, G., et al. 2019. Origin and fate of vanadium in the Hazeltine Creek catchment following the 2014 Mount Polley mine tailings spill in British Columbia, Canada. *Environmental Science and Technology* 53(8):4088–4098.

IMC (Imperial Metals Corporation). 2020. Mount Polley mine. www.imperialmetals.com/our-operations/mount-polley-mine/overview. Accessed September 2020.

McMahen, K., and Hughes, C. 2016. Mount Polley Mine August 2014 tailings dam embankment breach: Response and approach to remediation. In *The Technical and Research Committee on Reclamation, Symposium Proceedings.* Victoria, BC: Ministry of Energy, Mines and Petroleum Resources.

Miller, T., and Zapf-Gilje, R. 2016. *Mount Polley Mine Tailings Storage Facility, Perimeter Embankment Breach Update Report: Post-Event Environmental Impact Assessment.* Vancouver, BC: Golder Associates Ltd.

Miller, T., Zapf-Gilje, R., and Nikl, L. 2015. *Mount Polley Mining Corporation Post-Event Environmental Impact Assessment Report: Key Findings Report.* Vancouver, BC: Golder Associates Ltd.

Morgenstern, N.R., Vick, S.G., and van Zyl, D. 2015. *Independent Expert Engineering Investigation and Review Plan Report on Mount Polley Tailings Storage Facility Breach.* www.mountpolleyreviewpanel.ca/sites/default/files/report/ReportonMountPolleyTailingsStorageFacilityBreach.pdf. Accessed September 2020.

Lessons Learned from Tailings Dam Failures: A Social, Regulatory, Technical, and Governance Perspective

Harvey McLeod
Principal, Klohn Crippen Berger Ltd., Vancouver, British Columbia, Canada

INTRODUCTION

Tailings dam failures can cause significant environmental and social effects, while at the same time provide an impetus for step changes in human behavior, regulatory oversight, technical guidance, and governance for safer dams. Case history examples are used to illustrate these aspects and the corresponding key lessons learned from their failures. The potential consequences of a tailings dam failure vary widely from a limited loss of containment to a catastrophic loss, and the scientific basis for predicting the consequences continues to evolve. Public awareness with respect to the safety of tailings dams is driving toward greater transparency, and regulatory response to failures can drive toward clearer and/or more prescriptive controls on design and management of tailings dams (e.g., in Chile, Europe, Brazil, Canada). Technical and governance guidance continue to evolve to guide current and future designers and operators toward safer dams. The human element, both from the affected communities, as well as the design and operations sides, is explored.

REGULATORY LESSONS LEARNED

Regulatory learnings from tailings dam failures have normally been slow to occur. Tailings dams, unlike water dams, have traditionally fallen under a myriad of regulations that may be administered by water, environmental, or mining regulations, or combinations of regulations. Very few international regulations have been written specifically for tailings dams. In addition, regulators can be slow to react to changes because of variable public pressure and the process involved in achieving alignment between ministries.

El Cobre No. 1

The 1965 failure of the El Cobre No. 1 dam in Chile is an infamous tailings dam failure. It was the impetus for Chile to ban upstream dams, which was promulgated with allowance for exceptions in 1970, although it was not banned completely until 2007 (Lyu et al. 2019; Villavicencio et al. 2014).

Mufulira No. 3

In Zambia in 1970, the Mufulira No. 3 dam failed into underlying mine workings, which resulted in Zambia banning mining under tailings facilities (Regulation 602) (Ndulo 1986; Zambia 2008a, 2008b, 2014).

Mount Polley Mine

The failure of the Mount Polley mine tailings dam in Canada in 2014 initiated a review of international regulations on tailings dams, which concluded that there were no regulations, anywhere, that adequately addressed good governance and design for tailings facilities. The new regulations have more than 60 requirements for tailings dams and an associated guidance document (British Columbia, n.d.-a, n.d.-b).

Fundão and Brumadinho

The catastrophic failures of Fundão (2015) and Brumadinho (2019) demonstrated the uncertainty with upstream tailings dams and initiated two phases of changes in Brazilian legislation for tailings dams. A key learning, however, is that the regulations do not adequately address the mechanism of undrained strength and liquefaction behavior of tailings (BWI 2019; Palu and Julien 2019).

GOVERNANCE LESSONS LEARNED

Governance practices for tailings facilities have been slow to react to dam failures. Initially, practices strongly focused on health, safety, and environment, with the assumption that the safety of the tailings dam is under control by the engineers. As dam failures continued to occur in the 20th century, mine operators turned their focus toward good governance for tailings dams.

Omai and Aznalcollar

The failures of Omai in Guyana in 1995 and Aznalcollar in Spain in 1998 were the key impetus for the Mining Association of Canada (MAC) to develop guidelines and surveillance manuals for tailings dams (MAC 2019a, 2019b).

Fundão and Brumadinho

Fundão and Brumadinho were two catastrophic failures that changed the landscape with respect to governance of tailings facilities. The realization by mining companies that a dam failure could result in billions of dollars in damage and destroy the economic value of their company motivated mining companies and investors to attain a higher level of governance.

TECHNICAL LESSONS LEARNED

Technical learnings from dam failures include appreciation of the mechanisms of the dam failures, and, therefore, the advancement of technical knowledge. In addition, dam failures provide impetus for professional dam organizations, such as the International Commission on Large Dams (ICOLD) and their associated country organizations (U.S. Committee on Large Dams [USCOLD], Canadian Dam Association [CDA], Australian National Committee on Large Dams [ANCOLD], etc.) to develop technical guidance that builds on the learning and advancement in engineering practices (USCOLD 1994).

Omai, Aznalcollar, and Previous Failures

The period leading up to 2001 saw a continuous pattern of tailings dam failures that initiated technical organizations to develop professional practice guidelines. ICOLD (2001) and others created guidelines on mining and tailings dam safety (CDA 2007; ANCOLD 2019).

Mount Polley Mine

The dam failure at the Mount Polley mine initiated the guideline on Site Characterization of Dam Foundations in response to not only Mount Polley, but the many other tailings dams that failed because of poor understanding of the foundation conditions of the dam (Association of Professional Engineers and Geoscientists of British Columbia 2016; *Mount Polley Independent Expert Investigation and Review Report*, n.d.).

Fundão and Brumadinho

As described previously, the Fundão and Brumadinho failures also initiated changes in technical guidance for tailings dams. ICOLD (2001) has distributed a comprehensive guideline for safe governance and design of tailings dams. The guideline is currently being reviewed by member countries. Additionally, the Global Tailings Review (GTR) initiated global tailings standards, which include standards for governance and design of tailings facilities (GTR 2019, 2020).

SOCIAL LESSONS LEARNED

Social learnings from tailings dam failures can be less tangible, as most tailings dams do not fail, and failures, in some cases, impact a limited population. However, failure of a tailings dam can destroy families and the social infrastructure that relies on the mine, and it is necessary to be transparent with respect to the safety of a tailings dam and the potential consequences if the dam were to fail. Important learnings, both positive and negative, have emerged from the Brumadinho failure. On the "positive" side, the mine owners have a heightened awareness of the impact on the communities and the requirement to assure that these events do not happen again. On the negative side, fisheries in the river confluence with the Atlantic Ocean have been shut down because of misinformation on the toxicity of the tailings, which has negatively affected communities (Munro 2019).

REFERENCES

ANCOLD (Australian National Committee on Large Dams). 2019. *Guidelines on Tailings Dams: Planning, Design, Construction, Operations, and Closure*, revision 1. Hobart, TAS, Australia: Australian National Committee on Large Dams.

Association of Professional Engineers and Geoscientists of British Columbia. 2016. *Site Characterization for Dam Foundations in BC: APEGBC Professional Practice Guidelines. V1.2.* www.egbc.ca/getmedia/13381165-a596-48c2-bc31-2c7f89966d0d/2016_Site-Characterization-for-Dam-Foundations_WEB_V1-2.aspx. Accessed August 2020.

British Columbia. n.d.-a. BC mine information: Mount Polley mine overview. https://mines.nrs.gov.bc.ca/p/mount-polley/overview. Accessed September 2020.

British Columbia. n.d.-b. Spill management regulations. www2.gov.bc.ca/gov/content/environment/air-land-water/spills-environmental-emergencies/spill-preparedness-and-response-bc. Accessed September 2020.

BWI (Building and Wood Workers' International) 2019. New law guarantees more safety in dams in Brazil. www.bwint.org/cms/new-law-guarantees-more-safety-in-dams-in-brazil-1295. Accessed September 2020.

CDA (Canadian Dam Association). 2007. *Dam Safety Guidelines*. Toronto, ON: Canadian Dam Association.

GTR (Global Tailings Review). 2019. *Global Tailings Standard: Draft for Public Consultation.* https://globaltailingsreview.org/wp-content/uploads/2019/11/EN-Global-Tailings-Standard_CONSULTATION-DRAFT.pdf. Accessed November 2020.

GTR (Global Tailings Review). 2020. *Global Industry Standard on Tailings Management.* https://globaltailingsreview.org/wp-content/uploads/2020/08/global-industry-standard_EN.pdf. Accessed November 2020.

ICOLD (International Commission on Large Dams). 2001. Tailings Dams: Risk of Dangerous Occurrences: Lessons Learnt from Practical Experiences. Bulletin 121. Paris, France: International Commission on Large Dams.

Lyu, Z., Chai, J., Xu, Z., et al. 2019. A comprehensive review on reasons for tailings dam failures based on case history. *Hindawi Advances in Civil Engineering.* https://doi.org/10.1155/2019/4159306.

MAC (Mining Association of Canada). 2019a. *Developing an Operation, Maintenance, and Surveillance Manual for Tailings and Water Management Facilities*, 2nd ed. Ottawa, ON: Mining Association of Canada.

MAC (Mining Association of Canada). 2019b. *A Guide to the Management of Tailings Facilities*. Ottawa, ON: Mining Association of Canada.

Mount Polley Independent Expert Investigation and Review Report. n.d. www.mountpolleyreviewpanel.ca/. Accessed September 2020.

Munro, N. 2019. Jordan says conservatives spreading misinformation about liberal "plan to shut down" N.S. fisheries. *Chronicle Herald.* October 3. www.thechronicleherald.ca/federal-election/jordan-says-conservatives-spreading-misinformation-about-liberal-plan-to-shut-down-ns-fisheries-359032/. Accessed August 2020.

Ndulo, M. 1986. Mining legislation and mineral development in Zambia. *Cornell International Law Journal* 19(1), article 1. http://scholarship.law.cornell.edu/cilj/vol19/iss1/1. Accessed September 2020.

Palu, M.C., and Julien, P.Y. 2019. Review of tailings dam failures in Brazil. Abstract. www.researchgate.net/publication/337498009_REVIEW_OF_TAILINGS_DAM_FAILURES_IN_BRAZIL. Accessed September 2020.

USCOLD (U.S. Committee on Large Dams). 1994. *Tailings Dam Incidents*. Denver: U.S. Committee on Large Dams.

Villavicencio, G., Espinace, R., Palma, J., et al. 2014. Failures of sand tailings dams in a highly seismic country. *Canadian Geotechnical Journal* 51(4):449–464.

Zambia. 2008a. Mines and Minerals Development (General) Regulations, Statutory Instrument No. 84, commencing September 12. http://zambialii.org/zm/legislation/statutory-instrument/2008/84. Accessed August 2020.

Zambia. 2008b. Mines and Minerals Development (Prospecting, Mining and Mining of Uranium Ores and other Radioactive Mineral Ores) Regulations, commencing September 12. http://zambialii.org/zm/legislation/statutory-instrument/2008/85. Accessed August 2020.

Zambia. 2014. Mines and Minerals Development (Amendment) Act, commencing January 1, 2015. http://zambialii.org/node/10743. Accessed August 2020.

SESSION 2

State of Practice

CHAIRS

Amanda Adams, Stantec
Georgia Lysay, Freeport-McMoRan Inc.

PRESENTERS AND PANELISTS

Manuel Aparicio, Newmont
Ross N. Hunsaker, Newmont
Jim Obermeyer, Stantec
Clint Strachan, Stantec
Mike Waldron, Freeport-McMoRan Inc.

NOTETAKERS

Jim Casey, Klohn Crippen Berger Ltd.
Heather Lammers, Golder Associates/Colorado School of Mines

INTRODUCTION

The session on the state of practice focused on tailings management throughout the facility life cycle, with presentations and panel discussion on the design, construction, operation, and closure of tailings facilities. Presentations included case studies for facilities located in Guatemala, Mexico, Peru, and the United States. A common theme of close partnership between the design engineer, engineer of record (EoR), operator, and owner as key to successful design and effective project implementation continued through the design, construction, and operation presentations. Integrating closure planning during late-stage operations was the central theme of the closure presentation.

PANEL DISCUSSION: STATE OF PRACTICE

What changes have you seen in the state of practice during your experience?

In the 1970s, tailings management was in its infancy and TSFs were considered an afterthought and nuisance, rather than an integral component of the mine asset that must be engineered and properly managed as we see in today's practice. The industry has also evolved to include closure considerations in both TSF design and operation stages. Assurance processes have improved through better definition and designation of the EoR role and establishment of internal and external review boards for TSFs as an important element of tailings stewardship.

The rapid development of new technology in the past 40 years has had a major impact on the way TSFs are designed, operated, and managed. Modern tools, such as automated data collection and drone surveys, allow monitoring to occur on a near-real-time basis, which facilitates timely EoR recommendations. Access to data, software, and visualization tools also empowers industry experts in owner roles to do more technical work in-house and to not solely rely on external consultants.

In the industry, there is much room for improvement, including establishment of a clear and attractive career path for tailings practitioners. This effort is underway but must be accelerated to meet the current and future needs of the industry for tailings professionals. One way of doing this may be to encourage upper management and executives to visit TSFs more often and communicate with operations staff to better understand resource needs. Supporting staff and recognizing their contributions is key to retaining and building talented staff.

What indicators do you use to judge if a TSF is operating using good or leading practices? What may indicate poor practices?

It boils down to the people on the ground. Systems and practices that may be in place are only as good as the people implementing them. Communication between and engagement of operations staff is critical in all aspects of day-to-day operations. High staff morale and positive outlook are indicators that leading practices are being implemented. Engagement and clear interest from upper management in the TSF operation practices and staff are key to boost and maintain a positive operations team outlook and performance. Upper management should be engaging the TSF operations team and asking questions such as the following:

- Where are the challenges?
- What are the hardships?
- What are your resource needs?

How do you think the industry can do a better job of collaborating and communicating leading practices?

1. Develop tailings stewardship programs that promote collaboration between, and training for, practitioners.
2. Promote and facilitate sharing of successful practices and lessons learned between operators, managers, and engineers through visiting other sites and facilities.
3. Promote and facilitate sharing of information across the industry so that designers and operators can benchmark practices and share lessons learned.
4. Seek exchange opportunities to place consultants in owner operation teams to develop operational experience and to place members of the owner operation teams in consulting firms to develop analytical and design experience.

How does the role of review boards fit into the current state of practice? How does an engineer of record fit in?

Review boards have been common in the Canadian oil sands industry since the early 1970s, while the hard rock mining industry was slower to adopt the practice. Modern practice for many mining companies includes having both internal and external review boards. Terms of reference for review boards can widely vary, with some focused solely on the technical aspects of the TSF design and others focused on operations and organization aspects of tailings management. The commonly referenced Mining Association of Canada guidelines discuss review boards as a component of a project's assurance mechanism and as a critical control. It is crucial that the technical expertise of a review board be broad enough to cover the key technical aspects of a project (e.g., geotechnical, hydrogeology, etc.) and that board members with the appropriate expertise for the site conditions are engaged.

It is ideal for the EoR to also be the designer of the facility. The EoR tends to have the most institutional knowledge for the facility as it is common for there to be higher turnover of staff on the owner's side relative to the rare turnover of the EoR. The key components of the EoR's role during operations include monitoring the performance of the facility, reviewing instrumentation data, and continually assessing the safety of the structure.

The designation of EoRs and review boards has increased in popularity, and resource demand for both is exceeding supply. As an industry, it is vital we implement methods to develop our existing practitioners and train new staff to meet this demand.

What are mining companies doing to disrupt the status quo with respect to tailings state of practice?

Participating in symposia such as the Society for Mining, Metallurgy & Exploration (SME)'s Mine Tailings Symposium at the SME 2020 Annual Conference and Expo, collaborating with other mining companies, encouraging consultants to share information, and moving toward open and transparent communication of tailings management practices are all methods being used. Sharing successful practices and lessons learned is a step in the right direction.

Based on current challenges, what changes would you like to see in the state of practice for tailings in the next 10 years?

A clear need and demand for tailings curriculum to be developed and incorporated into the core curriculum at the existing mining universities exists, which may also be as focused topics at other education facilities. Industry needs to support this effort by engaging with universities to develop course content, provide lectures, raise awareness, and advocate for tailings management as an enticing career path.

Despite improvements in knowledge sharing, more work needs to be done. Mining companies should be open to hosting operators and/or consultants, who are working at other facilities, at their sites for knowledge sharing. More opportunities should be provided to teach mine operators the analytical tools used by consultants and engineers for them to gain a better understanding of how design and performance requirements influence operations for the facility. Working in silos has been identified as an issue that can be avoided in an organization with increased collaboration between work groups.

How can risk-based decisions and management systems be used to enhance tailings management?

Dashboard systems can be useful tools to monitor key performance indicators through data collection at the site level. These systems require thoughtful development during initial setup and some maintenance but are easy to implement in the field. The unique aspects of each site are considered during system development, and stakeholders need to understand what the system limitations and outputs mean to quickly make informed decisions. A benefit of the dashboard system is to proactively provide upper management an indication where potential issues can arise and where resources should be focused. FMI's dashboard tool has been embraced at all levels within the organization. It is important to have the EoR and operations staff engage in the development of tailings stewardship documents and manuals to maximize their benefit.

What has been the social impact of the Marlin mine closure? What lessons learned could be incorporated into closure planning for other projects?

Engaging the local community during the closure planning process and setting expectations for post-mining land use (e.g., farming and housing) is important to achieve a successful facility closure. Because employment at the mine decreases during post-closure, there is an opportunity to integrate post-closure land use and community employment possibilities. Options, such as declaring the area an ecological preserve, were considered at Marlin.

CLOSING THOUGHTS

In the state of practice session, mine operators and consultants provided insight on how tailings management practices have changed during their careers and ideas for how to drive the industry forward to meet current and future needs. The importance of collaboration across the industry was a key topic expressed throughout the session, with return benefits including the shared successes and lessons learned that can be incorporated into future designs and operation strategies. Panelists emphasized the critical roles of the tailings EoR and review boards, and the engagement of operations and management teams in tailings stewardship.

Cerro Verde Mine and Quebrada Linga Tailings Storage Facility: An Operational Excellence Partnership

Jim Obermeyer
Stantec, Denver, Colorado, United States

Amanda Adams
Stantec, Denver, Colorado, United States

Teresa Speigl
Freeport-McMoRan Inc., Arequipa, Peru

Tamara Johndrow
Freeport-McMoRan Inc., Phoenix, Arizona, United States

Shane Wiegand
Freeport-McMoRan Inc., Phoenix, Arizona, United States

INTRODUCTION

This paper summarizes the collaborative design and construction process of a new tailings storage facility (TSF) at the Cerro Verde mine that took place from 2008 to 2015. The close partnership between the designer/engineer of record (EoR) Stantec (formerly MWH), the mine owner and operator Sociedad Minera Cerro Verde S.A.A. (SMCV), and its majority shareholder, Freeport-McMoRan Inc. (FMI) facilitated a robust design and efficient project execution. FMI, SMCV, and Stantec continue to partner on the operational construction to rigorously achieve design expectations.

The Quebrada Linga tailings storage facility (Linga TSF) was a major component of the Cerro Verde production unit expansion (CVPUE) project. As a part of the CVPUE project, SMCV contracted Stantec to develop a design for Linga TSF to store 2.025 Gt (gigatons) of tailings generated at a rate of 240,000 t/d (metric tons per day). Linga TSF is located in Quebrada Linga, approximately 12 km south of the existing Quebrada Enlozada TSF (Enlozada TSF) and approximately 5 km south of the new concentrator plant.

Project Background

SMCV owns and operates the Cerro Verde mine, located approximately 30 km south of the city of Arequipa, in the District of Uchumayo, Province of Arequipa, Department of Arequipa, Peru. SMCV is majority owned by Freeport-McMoRan Inc., SMM Cerro Verde Netherlands (a subsidiary of Sumitomo Corporation), Compañía de Minas Buenaventura, and other minor shareholders including SMCV workers.

The Cerro Verde mine has been in operation since the early 1970s and is currently producing both copper and molybdenum concentrates and copper cathodes. When the CVPUE project was developed, operations consisted of two open pits, a solvent extraction and electrowinning (SX/EW) plant and associated heap leach facilities, a 120,000 t/d concentrator plant, and the associated Quebrada Enlozada TSF. The Enlozada TSF, which was commissioned in 2006, was designed for a total capacity of almost 1 Gt. Stantec was involved as the designer and EoR for the Enlozada TSF and has continued as the EoR during subsequent operations.

In June 2010, the new Linga TSF was selected as the best alternative for the CVPUE project. The selection of the site was followed by an embankment alternatives study that rated downstream method rock fill and centerline sand embankment construction as almost equal alternatives. Ultimately, the centerline sand embankment was identified as the preferred embankment type based on the excellent performance of the Enlozada TSF, water efficiency, and experience of the mine operators with this construction method. The feasibility study report was completed in March 2011. Basic engineering design began after completion of the feasibility report and included additional field investigations and analyses for Linga TSF. The basic engineering report was completed in December 2011. The final design phase was initiated in early 2012 and was completed in February 2013.

SMCV staff, with support and resources from FMI, had in-depth involvement and interaction throughout the design process. This collaboration included experts from environmental and permitting functions, site operations at the Enlozada TSF, the FMI corporate tailings and projects teams, as well as the EoR and on-site Stantec quality assurance (QA) and operations support staff. Close integration and coordination between the various stakeholders was maintained through the permitting, construction, and start-up phases.

Construction of the starter dam, seepage collection sump, and other capital construction items was initiated in May 2013 and substantially completed in September 2015 at which time Linga TSF was commissioned. Stantec provided construction engineering and QA and quality control (QC) and continues as EoR and provides QA and operations support during subsequent operation of the TSF.

Involvement of the external technical review board (review board) began during design of the Enlozada TSF. The review board's involvement continued through the design, construction, and start-up of Linga TSF and is still ongoing. During the design of Linga, presentations to the review board took place at key milestones and the review board made a minimum of one site visit per year.

Project Opportunities and Challenges

Collaboration between the on-site SMCV team, FMI corporate team, and Stantec EoR team was a key factor to identifying issues, developing solutions, and successfully implementing those solutions. Taking lessons learned directly from the existing Enlozada TSF design, performance, and operation, and applying them to the Linga design also contributed to the overall robust design that was achieved. The presence of a capable on-site SMCV operations team with experience building and operating similar dams directly influenced design decisions and was leveraged to achieve the compressed design, permitting, and construction schedule.

Key project challenges and the subsequent solutions developed by the team that are highlighted in this paper include the following:

- Production rate of 240,000 t/d, which must be stored in the TSF and has a short ramp-up schedule
- Water conservation and a water-efficient design for the TSF
- The ultimate dam height (305 m as measured at the dam centerline)
- The seismicity of the project site with interface magnitude (M) 9.0 and a peak ground acceleration (PGA) on rock of 0.32 g, and intraslab M 8.0 with a PGA on rock of 0.67 g
- The lack of readily available low-permeability material for use in the starter dam construction

The following sections describe the TSF design concept and provide an overview of the design and construction process, focusing on how the key challenges were addressed.

DESIGN CONCEPT

Linga TSF consists of a zoned rock fill starter dam that is raised by the centerline construction method, the tailings impoundment located upstream, and a seepage collection sump located downstream of the TSF embankment. Other components of the TSF are a two-stage cyclone station, reclaim water system, and tailings transport and delivery systems.

The 165-m-high zoned rock fill starter dam will be raised 140 m by constructing an embankment of compacted cycloned tailings sand using the centerline construction method. The ultimate height of Linga TSF will be 305 m at the embankment centerline and the tailings impoundment will cover an area of approximately 17,300 ha (hectares).

The starter dam was designed to provide storage for the cyclone tailings overflow and the portion of the whole tailings that are not cycloned during approximately the first two years of operation. This two-year period provided time for the compacted underflow sand embankment to be raised to a level slightly above the starter dam crest. This TSF embankment configuration was established to support the subsequent continuous raising of the dam with compacted cyclone underflow sand to provide sufficient freeboard for ongoing operation of the TSF during its 24-year design life.

A seepage collection system drainage network was designed to (1) enable the rapid drainage, transport, and recovery of water contained in the underflow sand; and (2) maintain a low phreatic surface in the starter dam and sand embankment during tailings deposition. The seepage collection system will be expanded as the embankment grows during ongoing TSF operations.

Linga TSF is designed and operated in a manner similar to that of the existing Enlozada TSF. The tailings slurry from the new concentrator is thickened to a concentration by weight (Cp) of 55%. The thickener underflow (whole tailings) is then transported through tailings delivery pipelines to Linga TSF. The material balance analysis demonstrated that 50% of the whole tailings materials needs to be cycloned at a two-stage cyclone station to provide sufficient quantities of cyclone sand to support ongoing construction of the TSF embankment, at a rate that maintains the 5-m design freeboard. The remaining 50% of the whole tailings is deposited in strategic locations to control the configuration of beaches in the TSF impoundment area to manage the location and size of the reclaim water pond. The produced cyclone underflow sand is deposited in 30-cm lifts on the 3.5:1 horizontal–vertical slope embankment and mechanically compacted. Water that drains from the hydraulically deposited tailings underflow sand is collected in a system of embankment underdrains, conveyed to the seepage collection sump, and pumped back to the cyclone station for reuse.

The main components of the initial capital construction included the starter dam, west tailings delivery corridor, two-stage cyclone station, seepage collection system (sump, embankment underdrains, seepage return water pipeline, and pumping system), and impoundment reclaim water system (pipelines, barges, and pumps). TSF construction also included jacking headers along the starter dam crest to support pipelines (cyclone underflow, cyclone overflow, and whole tailings) and to facilitate raising the pipelines as the dam crest is raised over time.

SITE CHARACTERISTICS

- Elevation: 2,300 to almost 3,000 m amsl (meters above mean sea level)
- Temperature: 0° to 30°C
- Average annual precipitation: 37 mm
- Evaporation: exceeds the precipitation by more than 60 times
- Probable maximum precipitation: 464 mm
- Probable maximum flood: 19,870,100 m^3
- Interface earthquake: magnitude (M) 9.0 with a PGA on rock of 0.32 g
- Intraslab earthquake: M 8.0 with a PGA on rock of 0.67 g

Geology

The geologic deposits at Linga TSF include volcanic ash, alluvium, and colluvium overlying metamorphic, volcanic, sedimentary, and intrusive igneous rock. Precambrian Charcani gneiss forms the basement of the region. It is exposed at the ground surface along a northwest trending belt (i.e., parallel to the strike of the Yura Group bedding and forms the exposed core axis of an anticlinal fold feature) across the middle portion of the impoundment area.

Hydrogeology

Groundwater flow in Quebrada Linga is generally convergent toward the main drainage channel, and the general flow direction is to the southwest along Quebrada Linga. The groundwater divides generally coincide with the location of the topographic basin divides between the Quebrada Linga and the surrounding valleys.

DESIGN HIGHLIGHTS

Design of the Linga TSF was accomplished by a multidiscipline team of Stantec design professionals and was coordinated closely with SMCV and FMI. The following sections discuss some of the design highlights.

Detailed Material Balance

The use of cyclone sand for TSF embankment construction required a detailed material balance to be developed for the design and maintained throughout project start-up and ongoing operations. A material balance model was developed early in the project and was subsequently updated through the design stages. The material balance analysis modeled the filling of the TSF impoundment and the raise of the embankment in monthly time steps throughout its operational life.

The density of the deposited overflow/whole tailings in the impoundment was estimated based on the actual results from the Enlozada TSF operation, large strain consolidation testing on representative samples, and analyses performed using the FSConsol computer code. The estimated density of the impounded tailings over time ranged from approximately 1.0 to 1.58 metric tons per cubic meter. Results of the material balance modeling were used to select the required cyclone sand production rate for use in design of the cyclone station and tailings distribution systems. The actual density is being updated throughout the operation, and the material balance is periodically updated to reflect actual conditions and to revise forecast conditions.

Detailed Water Balance

Because of the arid environment at this site, water conservation and a water-efficient design were priorities. Understanding the water losses required the development of a detailed water balance model for the TSF. The water balance model was directly linked to the material balance model, and like that model, used monthly time steps.

The main objective of the water balance analysis was to estimate the water losses from the tailings facility and the quantity of makeup water required for TSF operation. Results of the water balance analysis were used to establish the quantities of makeup water required throughout the life of the TSF for planning purposes.

A spreadsheet model was initially used for the design, but GoldSim software was later used for a site-wide water balance model that includes key site facilities such as the concentrators and tailings facilities, open pits, and heap leach facilities.

Extensive Geotechnical Investigations

The project team completed an extensive geological and geotechnical investigation and site characterization program that occurred in several phases. The program included detailed geologic mapping, geophysical surveys, test pits, and drill holes (both geotechnical and hydrogeological) conducted in a series of campaigns.

A total of 39 holes were drilled in the area of the Linga TSF project site during the feasibility investigation program (2008 and 2010). Fifty-three additional holes were drilled at the project site during the basic engineering and final design investigation programs (2011).

A total of 93 tests pits, road cuts, and stream channel cuts were logged and sampled between 2008 and 2011 within and around the proposed Linga TSF embankment and impoundment area as part of the geotechnical field investigation. Most of the test pits were used for borrow investigations.

In 2008 and 2011, geophysical surveys were performed at the location of the seepage collection sump and along the main embankment axis. During the final design phase, geophysical methods were used to supplement drill hole and test pit investigations at specific sites such as the rock quarries, seepage collection sump, and in areas of the main dam.

A thorough understanding of the embankment construction materials (primarily the underflow sand) was also required. The strength parameters used in modeling the strength of the underflow are based on the results of laboratory triaxial and direct simple shear testing performed on samples of underflow tailings. Consolidated-undrained triaxial testing and direct simple shear (DSS) testing was performed on remolded samples obtained from the Enlozada TSF. More sophisticated testing also included the following:

- Static triaxial testing at low to high confining pressures to develop shear strength parameters and evaluate effects of potential particle breakage
- Cyclic triaxial testing at low to high confining pressures to evaluate potential for liquefaction
- Cyclic DSS consolidated-undrained tests at low to high confining pressures to evaluate potential for liquefaction
- Shear wave velocity testing to estimate shear wave velocities with varying confining pressures
- Combined resonant column/cyclic torsional testing to develop modulus reduction and damping curves for the underflow

Static Deformation Modeling

The performance of the dam and in particular its central core significantly depends on the potential settlement and differential settlement that various portions of the dam experience. The potential for low confining stress zones at abutment contacts was evaluated to address the possibility for hydraulic fracturing of the core at abutment contacts and in the core itself. Three-dimensional static deformation modeling was performed using FLAC3D software (Itasca 2010). FLAC3D requires the generation of three-dimensional elements, which can be achieved through the use of two programs: Rhinoceros 3D and Kubrix. All three programs were used to develop and group elements in regions that allow for computational analysis.

Configurations for three stages of TSF development were modeled: starter dam at time of construction, starter dam and second-year embankment, and starter dam and ultimate height embankment. Actual deformation of the clay core was measured through the installation of four shape accel arrays (SAAs). The SAA motion was recorded and compared to the predicted values. The modeling and measured performance of the embankment since construction have shown that the predicted and actual deformations are within tolerable limits and were highest during the initial stages of operation.

Seismic Deformation Modeling

An analysis to estimate the potential earthquake-induced deformations of the TSF embankment at its ultimate configuration was performed using time-domain, two-dimensional, explicit finite difference code FLAC2D (Itasca 2008). The main objective of the analysis was to estimate the magnitude of potential earthquake-induced deformations. Deformations were then evaluated relative to freeboard, as excessive loss of freeboard can lead to uncontrolled release of the impounded tailings. Additionally, an unacceptable magnitude of deformation can lead to compromised internal features important to the performance of the embankment such as filters and drains. The results of the seismic deformation modeling indicated that expected deformations were within tolerable limits.

CONSTRUCTION HIGHLIGHTS

Successful completion of capital construction for Linga TSF required detailed planning and addressing key challenges. Selected construction highlights are discussed in the following sections.

Low-Permeability Core Material

The successful construction and operation of the starter dam required that 1.6 million cubic meters of low-permeability material be placed in the central core to meet the engineering design objectives. No significant sources of clay are present in overburden soils at the site.

Cerro Verde geologists noted the hydrothermal argillic alteration zones of the granodiorite porphyry within the leach cap rock positioned above the economic ore deposit in the western Santa Rosa pit. This material was the target of an investigation program by Stantec and SMCV that was developed and implemented to identify the quality and quantity of clayey material that could reasonably be harvested from the pit and the processing requirements to produce a clayey product with the required engineering properties for use in core construction. The most suitable material was relatively weak in situ, had a high clay alteration grade, and broke down when hydrated.

The altered zones were highly variable in their geological and engineering properties as well as spatial distribution throughout the pit. Considerable geological characterization was required in addition to coordination with the mine geology, short-term planning, and operations departments to identify and stockpile suitable leach cap rock material within the constraints and pace of an active mining operation.

Once the source material was identified, core zone material test fills were designed, observed, tested, and documented. These test fills confirmed the suitability and constructability of an identified source material in generating these materials during design of the starter dam. During construction, the SMCV short-range mine planning team and Stantec coordinated closely to identify specific target zones in the pit for core production. After suitability of the material was confirmed, SMCV adjusted the powder factor for the blast to increase material breakdown. The material was then transported to a stockpile where it was processed with power screens to remove particles larger than 15 cm.

The materials were still quite dry after screening and required further degradation into a more soil-like structure as well as moistening to near the optimum moisture content for compaction. The SMCV leach pad operators adapted irrigation systems like those used at the Cerro Verde leach pads to moisten the screened material in stockpiles near the starter dam construction area. This moistening was accomplished for a minimum period of one month, and preferably longer, to allow moisture to uniformly penetrate the stockpile and to provide sufficient time for the material to cure. The result was that the material broke down sufficiently to form a clayey, sandy gravel matrix that met the specification requirements for the starter dam core. In addition, the moisture content was very near the optimum moisture content for compaction when it was excavated from the stockpiles and hauled to the starter dam fill area. Only minor moisture adjustments were required prior to compaction. The in situ hydraulic conductivity of the core material after placement and compaction was within an acceptable range and the gradations and plasticity of the material met the specification requirements.

Construction Schedule

The required construction schedule was defined by the date that construction was permitted to start (May 2013) and date when the new concentrator commissioning was scheduled to occur (September 2015), but the starter dam construction had to be completed in August 2015 to allow time for jacking header construction and pipe installation on the starter dam crest by September. This provided a total of 27 months to complete activities for the starter dam, which included foundation excavation, grouting, material processing and production, and dam embankment fill placement. Some important factors to achieving the project schedule were

- A strong construction management team;
- Detailed planning and sequencing of all construction activities;

- Collaboration and cooperation between SMCV for self-performed construction activities with large mine equipment, FMI corporate for project support, and the EoR team for technical support;
- Use of large mining equipment capable of achieving required production rates;
- Use of two quarries to produce rock fill for the starter dam, one upstream and one downstream of the starter dam;
- Use of specialty contractors primarily for foundation grouting, aggregate material processing, and earthwork requiring smaller equipment; and
- Ability of SMCV to source additional mining equipment when necessary to maintain or recover schedule.

A simplified summary of the main construction quantities is presented in Table 1. The crest of the starter dam reached the design height of 165 m on August 8, 2015. The total volume of starter dam fill placed was 13,730,000 m^3.

Table 1 Starter dam construction quantities

Construction Activity	Quantity
Grouting	1,079 holes
Core production and placement	1,583,400 m^3
Filter production and placement	834,200 m^3
Drain production and placement	87,000 m^3
Transition zone production and placement	539,400 m^3
Rock fill production and placement	10,686,000 m^3

Courtesy of Stantec and Freeport-McMoRan Inc.

Starter Dam Construction Engineering and Quality Assurance/Quality Control During Construction

Construction engineering and QA/QC during construction were critical to confirm that the construction complied with design intent and construction specifications and to quickly respond to contractor questions and unanticipated conditions without delaying construction. Construction occurred 24 hours a day, 7 days per week, and resulted in on-time completion of the starter dam and other components of the TSF in time for the September 2015 concentrator commissioning date.

Stantec's primary responsibilities included providing field engineering and home office support services, observing the construction to confirm compliance with the construction drawings and technical specifications, addressing technical questions and issues that arose during the construction, conducting QA/QC, and field and laboratory geotechnical material testing.

Stantec had more than 75 on-site management and field personnel during the peak of construction.

START-UP AND CURRENT STATUS

Start-up of Linga TSF was challenging because of the high concentrator throughput, and also because the constrained valley in which the starter dam was located limited the surface area for placement and compaction of underflow sand for embankment construction. Approaches to address these challenges and the current status of the TSF are presented in the following sections.

Tailings Storage Facility Start-Up

Concentrator start-up for the CVPUE project began as planned on September 1, 2015. For the initial period, only whole tailings were deposited, as the cyclone station was not ready to produce sand for placement until October 20, 2015.

A very detailed start-up plan was developed by the EoR, FMI, and SMCV in advance of first tailings deposition. This detailed planning was key to the successful ramp-up of the project. The plan covered the two-year period until underflow placement was scheduled to reach the starter dam crest. This planning was critical because of the limited space available for underflow sand placement, which was the controlling factor in embankment construction rates during the early start-up period.

The start-up plan was effectively executed by experienced, capable, and diligent SMCV operations and management teams. As with all project start-ups, some challenges were encountered, but they were addressed in a timely manner by the SMCV and Stantec staff. Sand production and placement increased steadily from about 7,000 t/d at the start in 2015 to 29,000 t/d on average in mid-2016. Use of key performance indicators to monitor actual performance with planned performance was an important factor in the success of the first two years of operation and resulted in the compacted sand reaching the starter dam crest slightly ahead of time, on September 1, 2017, when the impounded tailings level in the TSF impoundment was 8 m below the crest. Figure 1 shows Linga TSF near the end of starter dam construction.

Tailings Storage Facility Current Status

In January 2020, the elevation of the sand embankment reached a height of 199 m at the embankment centerline at 2,604 m amsl, and the impoundment was at 2,596 m amsl. The height is measured from the base of the starter dam vertically to the lowest point of the embankment crest (the height measured from the downstream toe is greater than along the centerline of the embankment).

Courtesy of Stantec and Freeport-McMoRan Inc.

Figure 1 Linga tailings storage facility near the end of starter dam construction

Courtesy of Stantec and Freeport-McMoran Inc.

Figure 2 Linga tailings storage facility after four years of operation

Linga TSF is performing as anticipated in design and is considered to be highly successful. Linga TSF is shown in Figure 2 after four years of operation.

Inspections and oversight by the EoR are ongoing. QC is performed by SMCV, and QA activities and operations support are provided by Stantec 24/7. The FMI tailings stewardship team and the review board also annually inspect the facility.

At the most recent review board meeting, the board did not identify any concerns about overall dam stability. The instrumentation data indicate that the project is performing according to design expectations, and the review board noted that construction continues in accordance with good international practice.

REFERENCES

Itasca. 2008. *FLAC2D*, version 6.00 (revision 398). Minneapolis: Itasca Consulting Group, Inc.

Itasca. 2010. *FLAC3D*, version 5.00.101. Minneapolis: Itasca Consulting Group, Inc.

Construction and Operation of Tailings Storage Facilities: The Freeport-McMoRan Perspective

Mike Waldron
Freeport-McMoRan, Inc., Morenci, Arizona, United States

INTRODUCTION

Corporate affiliates Freeport-McMoRan Inc. (FMI) own and operate 16 active tailings storage facilities (TSFs) and manage 58 inactive or reclaimed TSFs in North America. This paper compares TSFs at two copper mines in Arizona: the Bagdad and Morenci mines. These TSFs are constructed utilizing upstream or centerline construction. The TSFs at the mines are located approximately 350 miles from each other. Variations in the history of development, processing infrastructure, and topography have led to varying scales of complexity and unique operating strategies at each mine.

TSF construction is unique and requires careful operation and oversight to apply best practice deposition, maintenance, and planning every day. Although both the Morenci and Bagdad mines have centerline cycloned tailings sand dams comprising their primary active TSFs, there are some noteworthy differences in operational strategies at these two sites. This paper explores these differences in the context of operational complexity, physical setting, and site requirements.

Strong corporate governance and collaboration within FMI result in the facilities being operated within the same culture of stewardship and managed under consistent systems. Successful operation requires ongoing checks and balances as represented in the FMI Tailings Management System (TMS). A selection of topics from the TMS are described in more detail and include the following: dedicated personnel with multi-tiered oversight; construction quality assurance (QA) and quality control (QC); key performance indicator (KPI) monitoring; regular site inspections; process documentation, and operator training.

BAGDAD AND MORENCI TAILINGS STORAGE FACILITIES' BACKGROUND

Following are brief descriptions of the two comparative mines and associated TSFs.

Bagdad Tailings Storage Facilities

The Bagdad mine is an open pit copper and molybdenum mining complex located about 100 miles northwest of Phoenix, Arizona. The concentrator at Bagdad processes approximately 82,000 short tons per day.

The Bagdad mine has two active and one inactive TSFs. The inactive TSF occasionally receives upset mill discharges into the impoundment. All three TSFs have been primarily constructed using the centerline method. The facilities were constructed using cyclones mounted uniformly along a steel header line that is supported and raised by a series of towers with hydraulic jacking systems. Figure 1 shows the centerline cyclone jacking header system operated at the active Upper Mammoth TSF.

Courtesy of Freeport-McMoRan Inc.

Figure 1 Bagdad mine centerline cyclone jacking header system used for tailings deposition

Morenci Tailings Storage Facilities

The Morenci mine is an open pit copper and molybdenum mining complex located about 200 miles east of Phoenix, Arizona. Morenci operations include two concentrators with a combined throughput of approximately 132,000 short tons per day.

The Morenci mine includes 10 TSFs: one active centerline facility, seven active upstream facilities, one inactive upstream facility, and one closed upstream facility (the status of active and inactive varies according to operating needs). Prior to mid-1980, Morenci's upstream TSFs were constructed utilizing traditional header and spigot methods. By the late 1980s, Morenci developed the use of crawler crane–mounted cyclone clusters to efficiently construct underflow sand shells to facilitate the upstream construction method as shown in Figure 2.

To support a mine expansion that included increasing mill production through a new concentrator, the West/East Dam (shown in Figure 3) was constructed as Morenci's first centerline TSF; it was commissioned in 2015.

Courtesy of Freeport-McMoRan Inc.

Figure 2 Morenci crane-mounted cyclone cluster on upstream tailings storage facility

Courtesy of Freeport-McMoRan Inc.

Figure 3 Morenci centerline tailings storage facility jacking header and cyclone system

OPERATIONAL DIFFERENCES

Because of the history of development, processing infrastructure, and topography, there are several noteworthy differences in TSF operational strategy and requirements between Bagdad and Morenci.

Overall System Complexity

Bagdad's current concentrator and the development of the TSFs began in 1977 with processing throughput generally consistent over the decades. Bagdad has operated one active primary TSF at a given time because of the mountainous topography consisting of large broad valleys and a consistent mill throughput.

The Morenci mill began operation in the early 1940s, and the milling throughput has varied over the decades because of metal price fluctuations and mill expansions. Areas around the Morenci mill consist of mountain terrain with relatively small incised valleys. For these reasons, along with evolving understanding of the need to manage raise rates on any single TSF, Morenci has constructed several TSFs and operates several at any given time.

The tailings distribution systems at Bagdad and Morenci differ based on the depositional requirements and the need to run multiple TSFs at a given time at Morenci. Bagdad's single tailings slurry pump station can deliver feed to three tailings delivery lines (TDLs). Morenci operates two slurry pump stations that can feed 10 TDLs through a complex network of pipes and valves to allow different pump trains to feed various dams.

Because of the topography and geology of a main valley fill embankment, Bagdad maintains two seepage ponds that dual perform as stormwater collection points for the TSF. Morenci, because of its operation of multiple TSFs and more undulating topography, maintains and operates well over two dozen seepage and stormwater facilities.

Operational Flexibility

Upper Mammoth TSF at Bagdad and West/East TSF at Morenci are similar centerline raise TSFs designed by the same engineer of record (EoR) consulting firm with similar jacking header tailings deposition systems. Because of differences in crest length and impoundment to embankment volumetric ratios, operating strategies vary between the facilities. Upper Mammoth TSF is an efficient structure requiring approximately 15% of tailings feed to be cycloned annually early in the life of the facility and reaching approximately 50% annually during the later operating years. A relatively low cyclone operating time in early years allows flexibility in system operation and in the deposition plan to facilitate maintenance, address instrumentation needs, and construct system extensions.

By comparison, because of the topography constraints and several existing TSFs within the footprint of the West/East TSF, its crest length is almost 3 miles with a crest line that quickly rises above natural topography. As a result, the West/East TSF mass balance requires a large underflow sand embankment volume compared to the impoundment volume. Morenci targets cycloning more than 90% of total tailings feed throughout the life of the TSF.

The Morenci team develops deposition plans as a best practice to allow adequate access and timing to perform maintenance activities, address instrumentation needs, and construct system extensions. The deposition plans are developed monthly and include the following considerations:

- Embankment slopes from topographic surveys—zones with flatter slopes receive lower tonnages than zones with steeper slopes.
- Jacking schedule—monthly jacking requirements are based on overall material balance. Short-term needs are driven by foot access to jacking pumps.
- Access to areas with instrumentation are maintained.

- Access to areas needing maintenance or system extensions are maintained.
- Impoundment geometry maintains reclaim pond location and proper beach width.

OPERATIONAL SIMILARITIES

Strong corporate governance and a culture of collaboration within the organization and with our EoRs result in the Bagdad and Morenci TSFs being operated and managed under consistent systems.

Internal Team Structure

The site operations team responsible for daily construction and management of a TSF includes key contributors, so comprehensive attention is paid to the various aspects that support safe and successful construction. These key contributors include knowledgeable and experienced operators, operational leadership with strong understanding of the drivers behind dam performance as well as potential risks, and strong on-site technical teams who routinely interact with TSF operators and associated activities.

Bagdad and Morenci operations have operational staff and frontline supervisors with decades of experience constructing TSFs at the respective mines. Systems and processes can always be improved, and continual improvement should be a philosophy of any industrial operation. In addition to being informed by good practices, EoR recommendations, and case histories, the current operational practices also represent an accumulation of lessons learned from years of managing tailings. Historical knowledge, including personal accounts, can quickly become an invaluable resource when making changes to a system or responding to upset conditions.

In recent years, the superintendents in charge of tailings operations at Morenci and Bagdad have well over a decade of experience with tailings dam design, construction, and operations. When daily operations are running smoothly, the operational team generally operates the TSFs as designed following training protocols. However, challenges resulting from upset conditions, swift changes in market conditions, or significant and sudden increases in operational demands require direction from a well-versed and experienced leader, such as those we have at Morenci and Bagdad. A leader who understands how decisions and actions impact short-term and long-term performance of the TSFs, has a strong understanding of their team's capabilities and limitations, and who readily uses internal engineers and the EoR for support, takes action that results in safe TSF performance and efficient utilization of resources.

Technical teams, such as an engineering group to support construction and operations of TSFs, can be structured and organized in several ways. Based on experience at FMI properties, technical team members who engage regularly with operations teams and regularly work near the TSFs are well positioned to provide guidance and oversight, maintain a high level of engagement with day-to-day activities, and develop practical solutions when needed. Strong collaboration, positive relationships, and frequent and direct communication between operational teams and engineering support have proved very beneficial at Bagdad and Morenci.

Engagement of Engineer of Record and Multi-Tiered Support

A successful approach at all FMI operations has been to engage personnel with different backgrounds, perspectives, and expertise to provide input into the system. The on-site TSF operations teams, including the engineering group, have several tiers of support and periodic reviews of operational performance and system designs.

One tier of support is the corporate team dedicated solely to tailings, crushed leach, and water (TCLW) functions throughout the organization. The TCLW team provides each site with a point of contact (PoC), typically an experienced professional chief engineer. The PoC is engaged in reviews of technical reports, design documents, as-built records, planning documents, and so forth, and participates in select inspections, as well as assists in coordinating corporate-wide initiatives, new technologies or practices, and obtaining resources across the organization. Each site also is assigned a director of tailings and water who provides oversight and technical direction, in addition to developing and sponsoring implementation of updates to the TMS.

Our experience is that a highly engaged EoR is essential for maintaining a high level of confidence in adhering to design principles and employing the observational method and provides another tier of support to the site. Mining operations and associated TSFs can be very dynamic and typically evolve over the long periods of time they are operated because of changes in society, technology, economics, mine plans, and/or regulation and also include turnover in staff. A consistent EoR can provide continuity to a facility if there are changes to operational staff. An EoR with direct access to historical information, design documents, and geotechnical data and who is routinely updated with operational status can provide the proper technical support to ensure that construction is in alignment with design intent. An EoR with an understanding of operations can often collaborate well with the operations team to develop safe and practical solutions that meet technical requirements.

Bagdad and Morenci maintain very good communications with their EoRs and have weekly (often daily) interactions. Design aspects or operating criteria, developed by the EoR, are reviewed by the on-site operations and engineering team and vice versa. For example, they review on-site personnel and provide feedback on EoR guidance on design of infrastructure, expansion plans, recommended operating criteria, or geotechnical investigation planning. On-site teams will request EoR review of in-house designs of stormwater management systems, installation of drainage measures, or deposition plans. The two-way review system reduces the potential for designs to be impractical or for poor performance of constructed infrastructure.

In addition to routine collaboration with the operations group, on-site engineering team, corporate PoC, other internal experts, and the EoR, FMI (via the Tailings Stewardship Program) employs third-party engineers to perform annual inspections and review of operations. FMI also facilitates regular engagements with technical review boards. These reviews represent a higher-level expert and external tier of support, and both are consistent practices at Bagdad and Morenci.

Quality Assurance, Quality Control, Key Performance Indicators, and Regular Inspections

Both Bagdad and Morenci have several programs in place to measure and document the quality of construction of the facilities. For capital TSF construction projects, a typical QA/QC program consistent with most construction industries is implemented. The primary form of QC currently employed at Morenci and Bagdad is routine sampling and laboratory testing of embankment materials as well as density testing when compaction is required. Routine quality testing does not negate the importance of regular comprehensive geotechnical investigations to adhere to the principles of the observational method.

KPIs have been defined at both sites and are regularly measured and reported on a monthly basis to the TCLW PoC and directors, the EoR, and on a quarterly basis to corporate executive leadership. Examples of KPI scoring parameters include piezometer levels and availability, beach width, freeboard, embankment underflow sand quality, raise rates, seepage flows, seepage clarity, and completeness of inspection-related action items. The KPI components and scoring criteria were developed through collaboration between corporate TCLW, the sites, and EoR and fall into the following categories: automated stability monitoring, reclaim pond management; dam construction/operation, and action item tracking for third-party recommendations.

Robust inspection programs are in place at Bagdad and Morenci and have been found to be effective in identifying conditions requiring monitoring or corrective action. In general, daily inspections are performed by operators; weekly and/or monthly inspections are performed by the on-site engineering team; monthly or quarterly inspections are performed by the EoR; and annual inspections are performed by the third party and the corporate tailings stewardship team. All inspections are documented and shared with appropriate personnel for review and to implement actions as needed.

Documented Processes

Although the preference of high-performing teams is to maintain experienced personnel to retain institutional knowledge and efficiency, turnover in staff is inevitable. The ability of an operational organization to maintain consistent practices through turnover in staff is a function of pre-planning and preparedness. One of the most reliable and impactful methods of maintaining continuity in task performance is the development and maintenance of documented processes.

Bagdad and Morenci have developed several documented processes. Some examples of documented processes include standard operating procedures, work instructions, operations and maintenance manuals, water operations manuals, written contingency plans, and the maintenance process manual. These documents not only help transition knowledge during staff turnover and during new employee training, but also become valuable references for retained staff.

Operator Training

A well-developed set of drawings, specifications, and operating criteria is useless without a dedicated and trained staff of operators and mechanics applying their skill sets and knowledge under guidance on a daily basis. Tailings dam operators generally are developed over time and with experience. Few opportunities exist to gain the required knowledge and experience to properly construct a tailings dam, other than functioning as a specialized tailings dam operator or mechanic for a period with targeted training and guidance from competent supervisors. Proper training is crucial to the development of supervisors and operators.

Bagdad and Morenci have defined lines of progression for operational and maintenance staff to establish career paths and expectations of competencies and functional tasks. New operators go through an established onboarding process at Bagdad, while Morenci's onboarding process currently is being redefined to ensure that operators are taught the basics of the systems and develop a general understanding of principles before learning more complex components or tasks. New operators work alongside experienced operators for a period, and they must demonstrate a base level of competency prior to solo operation of system components. Both sites employ a lead operator to work a straight day schedule to facilitate training of new employees and apply consistency across the typical rotating shift schedule.

In addition to normal task training and field training to perform expected roles, Morenci and Bagdad require operators to attend educational sessions facilitated by the EoR and site engineers. These sessions cover topics ranging from signs of abnormal conditions, potential failure modes, consequences of dam failures, technical background supporting defined operational criteria, utilization of operations and maintenance manuals, and implementation of emergency action plans.

CONCLUSION

Tailings dam construction is unique and requires careful operation and oversight to apply best practice deposition, maintenance, and planning every day. The comparison provided between Bagdad and Morenci is not to suggest either site has superior systems or teams in place. Rather, this comparison highlights that even TSFs with similar designs in similar climatic regions operated by the same organization can have very different operating requirements and strategies caused by differences in site history, topography, and site requirements.

Though there are noteworthy differences in TSF operations between the Bagdad and Morenci mine sites in Arizona, these TSFs are managed using consistent systems and frameworks. TSFs are constructed and operated through rigorous collaboration among various internal teams at FMI's corporate office and its subsidiaries (including site operations, site engineers, and corporate support) and external EoR for the investigation and site characterization, design, construction, safe operation, and monitoring of the TSFs.

Using Leading Practices to Disrupt Traditional Operation of a Tailings Storage Facility

Ross N. Hunsaker
Newmont, Zacatecas, Mexico

Ramon Arnaudez
Newmont, Zacatecas, Mexico

INTRODUCTION

For years, tailings planning and dam construction have been done using basic tools such as spreadsheets and Gantt chart software. Technology development has focused on the mining and processing functional areas, leaving behind the final resting spot for all the spent ore. The development of mine planning software has resulted in valuable tools that aid the mine planning cycle. However, these software tools have not been fully adapted to the planning and construction of tailings storage facilities (TSFs), thus leaving a huge gap in the mining industry. This gap must be closed.

The oil sand industry operates large and complex tailings storage facilities. Operators have taken advantage of the tools generated for mining and applied them to planning and scheduling for the construction and operations of tailings storage facilities. Several different software packages are required to handle this planning because of beach slope changes, mature fine tailings, and water management. Hard rock mining lags behind the oil sand industry when it comes to the use of mining technology applied to tailings planning, even though in many cases, planning conditions are not as complex.

Newmont's Peñasquito mine in Zacatecas, Mexico, operates a TSF with a perimeter of 11 km and a 5-m rate of rise. The TSF is a centerline raise for three sides of the dam and a downstream raise for the fourth side. The TSF is currently being constructed using a mine fleet 930E for a buttress, a separate fleet of Caterpillar (Cat) 777 haul trucks for a sliver fill, 20-metric ton dump trucks for rock fill, and cyclone sand.

The Peñasquito tailings team recently began implementing mine planning software to integrate all the different construction and operation activities into a single plan. Scenario planning has been greatly enhanced by using software to optimize resources, activity duration, and constraint identification. The overall implementation of the software remains a work in progress with the universal goal of a safe, stable, and well-planned storage facility.

SITE OVERVIEW

The Peñasquito polymetallic mine is the fifth largest silver mine in the world and the second largest in Mexico. It is located in the northeastern corner of the State of Zacatecas and is wholly owned by Newmont Corporation. Open pit mining began in March 2010.

As can be seen in Figure 1, Peñasquito's TSF is one of the largest tailings dams in the world with a perimeter of 11 km. It is a four-sided dam with the west, north, and south dams being built as centerline and the east dam as downstream. The ultimate capacity of the TSF is 809 million t (Figure 2).

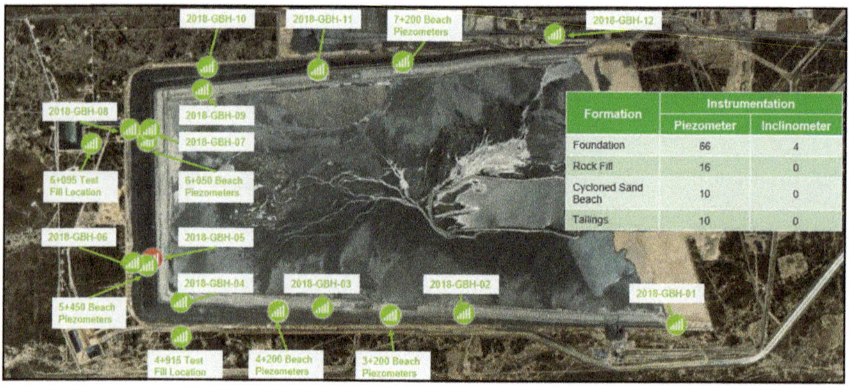

Courtesy of Newmont

Figure 1 Peñasquito tailings storage facility

LEADING PRACTICE: TAILINGS PLANNING AND CONSTRUCTION SEQUENCING

The multiple work fronts and equipment interactions make for complicated planning processes at Peñasquito (Figure 3). A fleet of 930 trucks are used for buttress placement, Cat 777 trucks for sliver fill, and 20-t trucks to place rock on the crown. Dozers and excavators are used for sand and rock placement.

Planning and scheduling activities for the different fleets of equipment consisted of three different planning groups: One group worked on the buttress and sliver fill planning, a second group oversaw rock placement and sand boxes construction, and a third group was dedicated to plan for sand

Key Design Criteria	
Tailings Facility	Peñasquito Tailings Storage Facility
Perimeter Length, km	11 km
Dam Raise Methodology	Centerline (North, South, and West) Downstream (East)
Maximum Dam Height, m	100 m
Tailings Storage Capacity, Mt	809 Mt
Tailings Processing, t/d	110 kt/d

Courtesy of Newmont

Figure 2 Key design criteria and capacity

Dam Construction Equipment

- 930s (Mine)
- 777s (24)
- 14m^3 Trucks (30)
- Backhoes (6)
- Dozers (4)
- Water Trucks (2)
- Grade (1)
- Smooth Drum Roller (1)

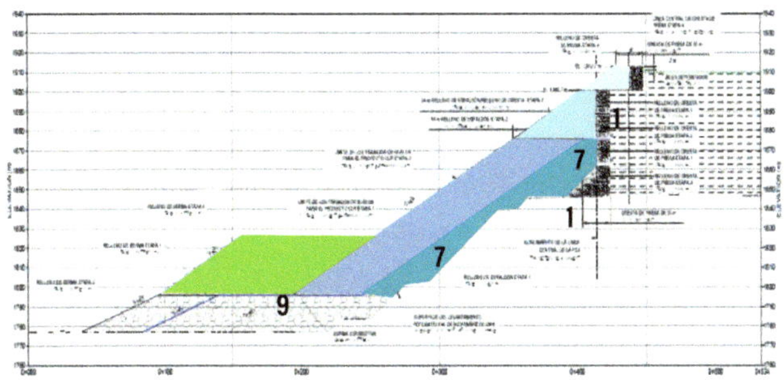

Courtesy of Newmont

Figure 3 Construction equipment planning

placement (Figure 4). Having different groups created gaps or lack of coordination between activities. To solve this issue, Peñasquito began evaluating mining application tools suitable for tailings dam construction scenarios. The Deswik suite of software (2020), was used to schedule mine waste from the open pit and planning for its placement in the embankment. The interface between designing, scheduling, and construction was facilitated by three-dimensional (3-D) visualization. The model allowed the planning team to create a fully integrated tailings plan, managed by one group instead of three, helping to eliminate the gaps or misalignment between groups.

Courtesy of Newmont

Figure 4 View of the embankment showing the three zones considered for planning

The crown fill placement alone involves six different tasks, each one requiring a specific set of equipment. For example, while rock placement only requires dozer work, pipe berm construction requires both backhoe and dozer work. Even though only one piece of equipment is assigned as "operating," the other can be assigned as required for hour calculations. Figure 5 shows a typical cross section with the different activities involved.

The overall tailings model is composed of three main components as shown in Figure 6. The mass balance model feeds the Deswik model with the quantity of generated underflow sand and the total amount of tailings placed in the dam. The rock placement plan is fully generated within the Deswik application, and it is driven by loading and hauling capacity.

Figure 7 shows a screen capture of a small section of the TSF from the Deswik model. A backhoe is shown working on the pipe berm construction while the dozer works on sand placement.

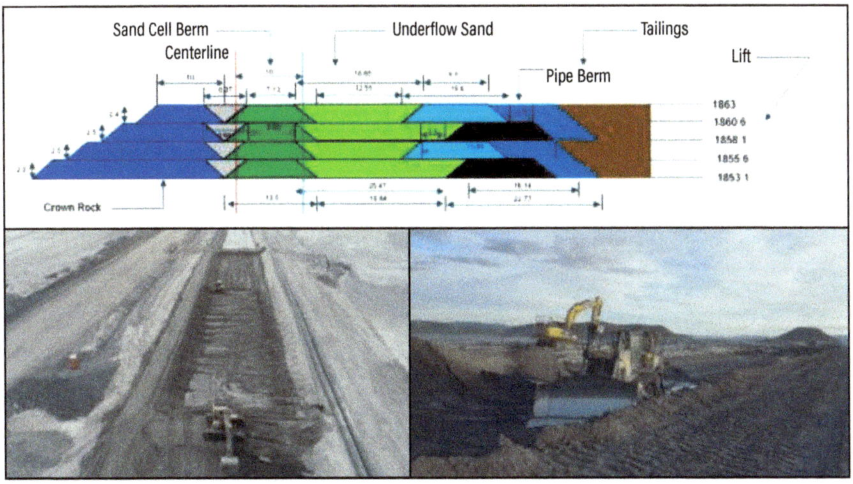

Courtesy of Newmont

Figure 5 Tailings dam cross sections and construction

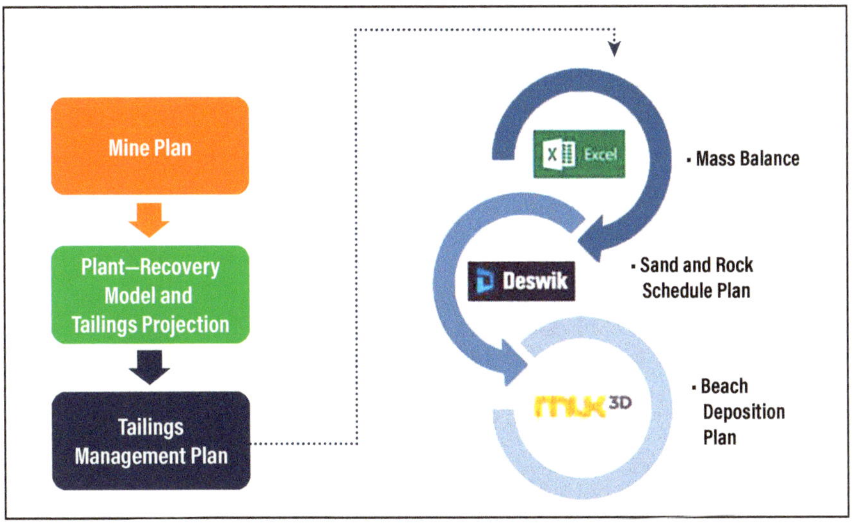

Courtesy of Newmont

Figure 6 Mass balance model

A broader picture of the entire TSF is shown in Figure 8 where each color represents a specific activity.

USING MODERN TECHNOLOGY TO COLLECT ACTIONABLE INFORMATION

Having 10 km of beach perimeter and a 5-m rate of rise produces challenges such as maintaining suitable beach lengths across the dam. In the past, spigot

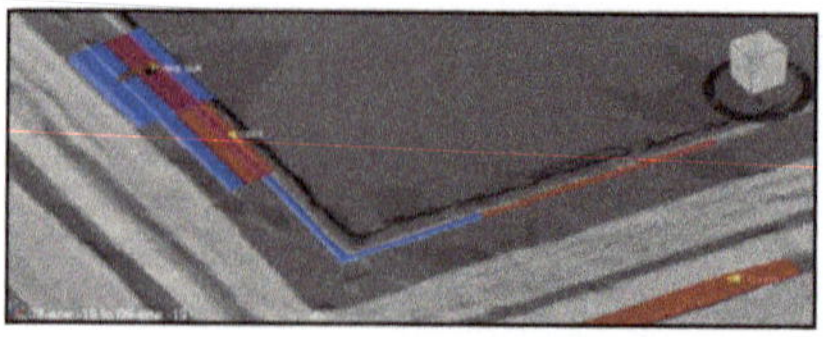

Courtesy of Newmont

Figure 7 Tailings storage facility model showing pipe berm and sand placement

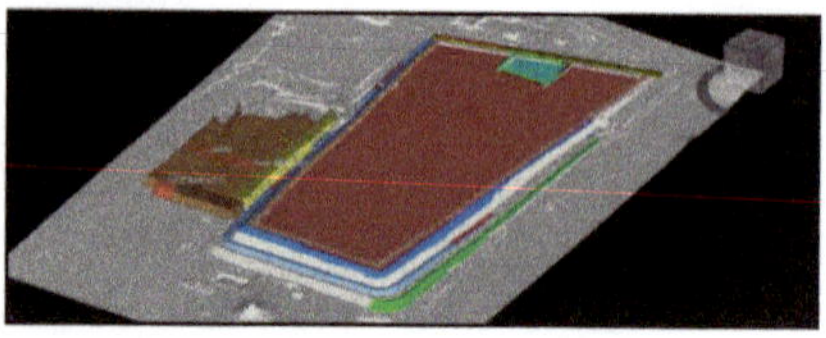

Courtesy of Newmont

Figure 8 Tailings storage facility model showing color-coded activities

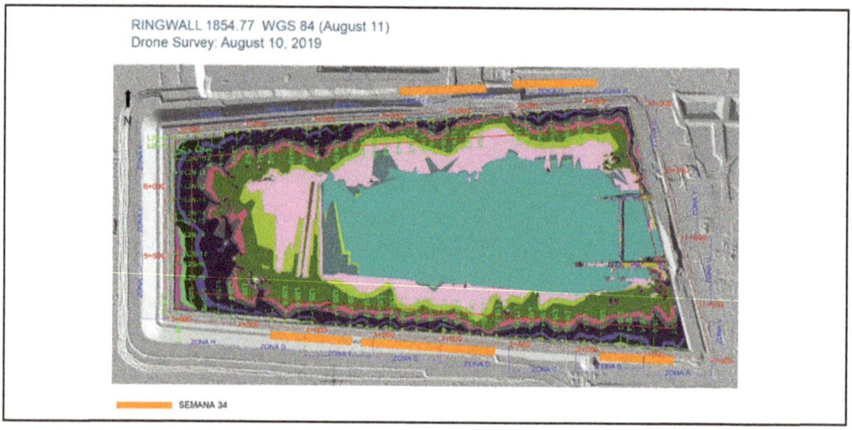

Courtesy of Newmont

Figure 9 Weekly drone survey output

discharge was planned based on visual inspection from operators and supervisors, which create discrepancies between crews and shifts. To eliminate inconsistencies and create a reliable beach management plan, a weekly drone beach survey was scheduled (Figure 9). Cloud data from the drone were uploaded into the Deswik software application to create a colored elevation model that clearly allows operators and supervisors to identify potential low points at the beach. A deposition plan is provided to the operations team to maintain beach compliance.

INTEGRATION OF GEOTECHNICAL INSTRUMENTATION NETWORK AND THE SITE TRIGGER ACTION RESPONSE PLAN

With the unfortunate tailings dam failures in recent years, having conventional (manual) instrumentation is just not enough. In 2018, Peñasquito switched from conventional monitoring to a telemetric system, allowing the geotechnical engineer to invest more time on data analysis than field data collection. Their system, provided by Sensemetrics (2020), is monitored in an on-site control room. Piezometers and inclinometers are linked to an alert system based on Peñasquito's trigger action response plan (TARP). In the event that a delta pressure exceeds limits defined in the TARP, alerts are automatically sent to key personnel.

UNITING THE TECHNICAL, COMMERCIAL, AND SOCIAL DOMAINS OF WORK

If we accept that work is turning intention into reality and that a company's organization is a social construction (Macdonald et al. 2018), then to be successful, it is important to have clarity on task assignments, levels of authority, and expectations. In mining, there are systems around technical work (e.g., the way a company operates) and commercial work (how the company sustainably exists). However, there seems to be a lack of social systems (how people interact, how tasks are assigned, etc.). For example, would you ask an entry-level operator at a tailings facility to conduct dam breach analysis of the tailings facility as his or her first job task? The answer is no; however, "we see in many instances people put into roles which have significant work in the social domain; such as large leadership roles, with little or no clarity of expectation or training" (Macdonald et al. 2018).

Without proper balance between these work domains (Figure 10) and strong systems to support them, the likelihood of success in turning intention into reality in the workplace is greatly reduced.

Implementing mine planning software to integrate all the different construction and operation activities into a single plan has been very helpful to Peñasquito mine development. The overall implementation of the software, with the universal goal of a safe, stable, and well-planned storage facility, remains a work in progress, and the addition of scenario planning has resulted in optimization of resources, activity duration, and constraint identification.

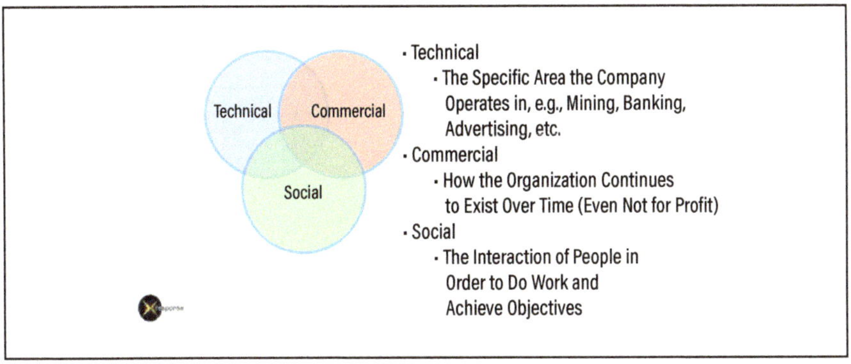

Source: Macdonald et al. 2018, © MacDonald Associates Consultancy Ltd.

Figure 10 Technical, commercial, and social domains in the workplace

This progress allows the mine to improve social systems' increase by focusing on clear communications and expectations.

REFERENCES

Deswik. 2020. Software suite. www.deswik.com/software/. Accessed June 2020.

Macdonald, I., Burke, C., and Stewart, K. 2018. *Systems Leadership: Creating Positive Organizations*, 2nd ed. New York: Routledge.

Sensemetrics. 2020. Software suite. https://sensemetrics.com/platform/software/. Accessed June 2020.

Case Study: Pre-Closure Tailings Management

Clint Strachan
Stantec, Fort Collins, Colorado, United States

Manuel Aparicio
Newmont, Guatemala City, Guatemala

INTRODUCTION

This paper outlines the closure planning and pre-closure work conducted at a precious metal mine in Central America. Mine operations included open pit and underground mining and a conventional mill. The final years of mine operation included tailings disposal with both conventional slurry and pressure filter plant operations with focused site water management in a climate with distinct wet and dry seasons.

Tailings storage facility closure was aided by closure planning and pre-closure construction and operation. This included construction of a spillway designed for tailings and water management during the final years of operation as well as the capability to discharge extreme-storm runoff during operations and post-closure. The final years of slurry tailings deposition followed plans developed to create a draining tailings surface amenable for cover placement and establishment of vegetation. The open pit mine was partially backfilled with filtered tailings and underground waste rock to enhance slope stability and provide geochemical isolation of the pit walls. Reclamation cover materials, selected plant species, and vegetation establishment techniques were chosen and tested at various locations on-site while conducting concurrent reclamation during operations.

This paper summarizes lessons learned as well as the successes in this pre-closure and closure process.

SITE CONDITIONS

The site (approximately 120 km west-northwest of Guatemala City) is in a mountainous region of western Guatemala at elevations ranging from approximately 1,600 to 2,500 m. Surrounding area land use includes livestock grazing and small-scale farming.

The site is in an area of high seismic activity, influenced by strike-slip faulting and subduction zone activity from a regional fault system that separates the Caribbean, North American, and Central American plates. Site bedrock units are typically Quaternary pyroclastic deposits.

Long-term (post-closure) conditions were analyzed for ground motion from the maximum credible earthquake (MCE). In the site area, the MCE is a Richter magnitude 8.9 event, which can generate an estimated peak ground acceleration in rock of 0.51 g (or 51% of the acceleration of gravity) at the site.

The regional climate reflects distinct wet and dry seasons. The average annual precipitation is approximately 1,200 mm, with most precipitation occurring during a bimodal wet season, typically from May through October. Annual pan evaporation averages approximately 1,700 mm. Average maximum and minimum temperatures (measured at Huehuetenango) are 25°C and 10°C, with an average annual temperature of 18°C.

MINE BACKGROUND

Following permitting and site development, mining and milling operations were started in late 2005, using both open pit and underground mining methods. Open pit mining was phased out in 2012, while underground mining continued to the end of milling operations in 2017. Processing for precious metal recovery utilized conventional milling, with a cyanide leach process and Merrill-Crowe recovery circuit for recovery of gold and silver as doré. Mill production ranged from approximately 4,000 to 6,000 t (metric tons) per day, with overall mine production at approximately 250,000 equivalent gold ounces per year.

TAILINGS MANAGEMENT

The tailings storage facility (TSF) was located in a valley downstream from the mill and designed and operated to store mill tailings and process water. Containment was provided by the underlying volcanic rock and a cross-valley dam (main dam) and extension along a ridge (saddle dam). The TSF dams were designed as water storage dams, with low-permeability clay cores, rock fill shells, and filter and drain zones. The clay cores included key trenches excavated into the foundation, which was supplemented with grout curtains and caps where necessary. The dams were constructed in phases in a downstream manner, with the maximum height of the main dam being approximately 88 m.

Design and operation of the TSF were in compliance with applicable laws pertinent to mining operations (Law Decrees by the Congress of the

Republic of Guatemala), applicable guidelines (from International Finance Corporation, International Commission on Large Dams, and Mining Association of Canada), and Cyanide Code requirements for the transport, receipt, storage, use, and decommissioning of cyanide. Based on potential downstream impacts, the main dam was classified during operations as a high-risk or high-hazard structure.

For TSF water management, mill process water, underground mine water, and precipitation runoff were retained in the TSF for mill makeup water in the dry season. Excess water in the TSF could be discharged if Guatemalan Ministry of the Environment (MARN) discharge water-quality standards were met.

Mill tailings were discharged into the TSF as a slurry (at approximately 35% solids). The average tailings density in the TSF increased from approximately 0.7 t/m^3 at mill start-up to approximately 1.3 t/m^3 at the mill shutdown. Depending on the volume of water stored in the TSF, tailings discharge was either subaerial or subaqueous.

PRE-CLOSURE MODIFICATIONS

Based on additional ore reserves, the crest elevation for both the main dam and saddle dam was raised from the original design elevation of 1,954 m to a final elevation of 1,962.5 m. However, the capacity provided by the additional dam raises did not meet life-of-mine tailings capacity requirements. Design of a tailings filtration plant and associated tailings delivery system was initiated in 2010, with the system commissioned in 2012. Filtered tailings were placed as partial backfill in the open pit mine, with the TSF as a secondary disposal location for tailings slurry.

With the TSF dam crest elevations established, a spillway was designed to accommodate operating and post-closure conditions. The spillway was constructed in 2011–2013. The spillway included a labyrinth wall at the invert, with sluice gates at different elevations to allow permitted discharge of water from the TSF at varying pool elevations. The spillway wall also provided accommodation for the tailings surface elevation at closure because the final volume of tailings to be discharged in the TSF was not established at the time of spillway construction.

After 2012, tailings were filtered, conveyed, and trucked to the open pit for partial backfill or discharged as a slurry in the TSF. The slurry discharge was enhanced by addition of discharge pipelines and spigots around the perimeter of the TSF, as well as addition of a floating discharge line. This allowed development of a tailings beach consistent with a post-closure surface that would drain to the spillway. When available, tailings were also spigoted in areas of the TSF with the deepest tailings to compensate for expected future settlement in those areas.

TSF water management was enhanced in later stages of operation with installation of a water treatment plant. Although the mine is in a region where average evaporation exceeds average precipitation, discharge of underground mine water into the TSF created a net gain of water, requiring treatment and discharge of excess water during the dry season. Water treatment plant operation also enhanced closure scheduling by reducing the volume of process water in the TSF at closure and allowing tailings beach surfaces to dry out and gain shear strength.

CLOSURE ELEMENTS

A comprehensive mine site closure plan was prepared as part of initial mine permitting. The closure plan was updated several times during mine operation as site conditions changed. The most recent closure plan was updated near the end of mill operations and has been modified during closure to coincide with site conditions. Many of the key closure elements have benefitted from observation and planning prior to closure, as summarized in the following paragraphs.

The TSF cover surface has been designed and constructed as a draining surface that discharges to the spillway. The cover material has been excavated from nearby borrow areas, with the cover thickness and characteristics based on separation from underlying tailings, placement method and equipment, and root depth of preferred vegetation. The vegetation established on the cover surface is based on knowledge of preferred species in the region and establishing a full and self-sustaining group of plant species that provide erosion protection for the cover and discourage invasive species.

Although most of the topography upstream from the TSF originally drained onto the TSF, some diversions were constructed during operation to improve the water balance. Post-closure site drainage includes diversion channels and regrading where possible to direct runoff away from or around the TSF. Remaining areas that cannot be diverted are directed onto and over the TSF cover in channels designed to convey extreme-storm runoff without eroding the TSF cover. These channels route the runoff over the cover surface through the spillway. Drainage planning was enhanced by observation of erosion and sedimentation on site, as well as performance of diversion and drainage channels constructed during operations.

The performance of the TSF dams has been monitored with piezometers, survey monuments, accelerometers, and nearby groundwater monitoring wells. Piezometers and survey monuments have been replaced as equipment has ceased to function or been damaged. Post-closure planning includes installing monitoring equipment designed for long-term operation, automated data collection, and remote data review.

CONCLUSION

Closure of the mine site has been aided by concurrent reclamation and pre-closure observation, testing, and planning. Other site conditions, design criteria, and operational elements that aided in closure are summarized:

- Guatemala is a country relatively new to industrial mining; the design criteria, design analyses, and construction quality assurance were consistent with international standards.
- Independent technical review was utilized throughout design, construction, operation, and closure planning. This helped to confirm or improve some of the key design and construction components.
- The potential for acid generation of mined materials was identified prior to and during mining, with proactive waste management handling and encapsulation plans.
- Permitted discharge of water from the TSF was carefully controlled, utilizing TSF water management and active water treatment.
- In later years of mine operation, the remaining life-of-mine was known, and periodic deposition plans were developed, allowing more accurate closure planning for the TSF and other mine facilities.
- Mine areas were available for concurrent reclamation during operations, which allowed observation of reclamation performance for vegetation species selection, erosion control materials and placement techniques, and sediment control.
- Updates of the closure plan during operations allowed improvements to be made for changing site conditions and revised operating plans. Updates to closure elements during actual closure work allowed improvements and cost saving to be realized.

SESSION 3

Tailings Governance

CHAIRS

Harvey McLeod, Klohn Crippen Berger Ltd.
Caius Priscu, Anglo American

PRESENTERS AND PANELISTS

D. Chad LePoudre, BHP
Kimberly (Finke) Morrison, Newmont
Caius Priscu, Anglo American
Teresa Speigl, Freeport-McMoRan Inc.

NOTETAKERS

Kelly Ward, Marsh
Christina Winckler, AECOM

INTRODUCTION

The goal of this session was to understand various governance systems—either those currently in place or being developed—and their advantages and limitations. We define governance as a culture of performance that works throughout the organization with a core value of reducing risk. Governance systems start in the corporate office, and by necessity, they are a mix of managerial, technical, and operational components.

Presentations in this session included corporate governance models for tailings storage facilities that have been implemented to reduce the risk to owners, shareholders, and the public, and technical and operational governance systems and components.

PANEL DISCUSSION: TAILINGS GOVERNANCE

The Tailings Governance panel discussion was lively, and responses were not universal among the panelists, with wide-ranging differences in company policy.

What is the auditing process at the corporate level and what kind of systems are in place?

Corporate teams are executing a review of the standard implementation and critical controls at the sites using independent auditors together with risk engineers. Typically, all the site reviews are completed within a two-year cycle.

New essential function review has been instituted that audits the protocols and processes at the mine site level. An outside facilitator functions as an auditor and asks questions in an interview style of the site personnel. Each site is reviewed for documentation conformity, and good practices are identified along with staffing needs.

Recently developed key performance indicators (KPIs), also called *key risk indicators*, are reviewed and used to assess each site. The reports on failures (e.g., Mount Polly, Brumadinho, Samarco) are used by the sites to rank and/or score against the identified "contributing factors."

New audit programs are being developed. Priority rankings are starting with timelines and responsibility from the sites to follow up with corporate offices. Nonconformance must be explained.

Regarding training programs, how are they set up and are they standardized?

In many situations, it is in the scope of work for the engineer of record (EoR) to train site staff members. The corporate team is responsible for identifying leading practices and distributing lessons learned from past incidents. A competent person has been assigned at each site. Training is an annual budget item, and it is executed as an annual session. In some cases, this type of training is a mandatory activity for key staff members.

Several technical trainings for mine site and corporate engineers are offered both internally and externally. Annual summits with engineers from all mine sites are held to share lessons learned and outcomes of corporate initiatives. To have consistent training throughout the sites, there is a plan to standardize the operator training.

Finally, there is an identified need to initiate certification and training programs with different universities and room for collaboration to get competent persons the right training.

What about the stakeholder role and engagement?

The stakeholders can ask questions during stakeholder days and visit the sites, which also serve as an educational opportunity. Communities have opportunities to visit sites with the site managers. When there is a new project or expansion, the social aspect is included in evaluations, and stakeholder engagement is both needed and encouraged in the planning phase.

Gaps have been identified, especially regarding stakeholder engagement with inundation studies and emergency action plans (EAPs). New standards are being written that directly address this engagement and how it can be executed. Sites within the portfolio are being identified that do not have stakeholder engagement, and these are prioritized. Peñasquito's engagement with stakeholders is a good example to follow with EAP tabletops, site tours, education about construction methodologies and evaluations, and dedicated muster points.

The key to acquiring and maintaining the social license to operate is stakeholder engagement. Mining companies have learned more about this post-disaster, and rather than preventing process, companies have then dived deep into governance processes.

What are the companies thinking about climate adaptations and seismic activations? How are these aligned with sustainable development? Should tailings dams be approached as "process safety" rather than geotechnical facilities?

Tailings dams are managed in a process safety approach with regards to looking at the dams holistically and on multiple levels using KPIs or critical controls. For example, the reclaim ponds are monitored against operational criteria, and inspections and resulting action items are tracked. Before changes are implemented on a site level that could affect safety, a corporate team reviews and engages appropriate stakeholders.

Sensitivity analyses are being completed to simulate dry and wet year cycles at all facilities and to evaluate severe climate impacts. Sensitivity analyses are also performed regarding the quality of hydrologic data. Seismic hazards maps are being updated in South Africa in partnership with the South African Council for Geoscience and several other partner institutions, universities, and research organizations within South Africa, with plans to make them public in 2020. Independent reviewers are hired to review the seismic hazard maps.

A consultant has been hired to provide a 40-year look ahead to identify the most impacted sites for detailed planning. Starting this year, environmental teams plan to hold climate change workshops at the sites. In addition, the technical review boards for both inactive and active sites have hydrology and seismic experts, and their recommendations and findings are taken seriously and followed closely.

What do you make of intentionally separating the engineer of record from the designer of record?

The Anglo American technical standard available to download on the company website (www.angloamerican.com/sustainability/tailings-summary/managing-tailings-safely) specifically states the EoR is—preferably—the designer of record (DoR) and that continuity is important and needs to be maintained. If the DoR cannot move forward to EoR, for a variety of reasons, a significant and detailed change management plan is required, which is *not* the preferred methodology nor an ideal strategy.

Some sites within the portfolio are in transition with the EoR and DoR as separate entities. However, everyone acknowledges the intent of transitioning to a single firm. And situations in which firms lose individual practitioners need to be monitored and addressed appropriately. The capability of the firm needs to be re-evaluated as senior practitioners leave, change employers, or retire.

Are there any components of pending regulations or the *Global Industry Standard on Tailings Management* that you have concerns about implementing?

Regarding pending regulations, there is not a lot of guidance from the Brazilian government where the threshold or bar is with respect to decharacterization of tailings dams built using the upstream method. Furthermore, there was consensus among the panelists that the International Council on Mining and Metals (ICMM) companies are ready to implement and follow the *Global Industry Standard on Tailings Management* (GISTM). The step change is necessary for the industry. It was also identified that smaller mining companies might have challenges to implement the GISTM because of resource needs. However, this should not be a reason for not implementing leading practices at all sites.

Regarding artificial intelligence, what is the value in digital tools and digital twins to monitor compliance and where does your company stand with this?

Artificial intelligence is probably most applicable to mechanical equipment, for example, optimizing a cyclone station, monitoring pumping systems and gates, and for spillway control. Although these tools are important, we can never replace the human element and good judgment. We need to be careful with the digital twins buzz word. The tools are especially helpful for collecting and managing massive amounts of data and can help us recognize when improvement is needed. However, we need to see how the tools work together and across engineering specialities. In the next 5 to 10 years, we foresee a big evolution in the use of these tools and their capabilities, and the opportunity lies in learning how to leverage that. Applying artificial intelligence solutions to earth science, geoscience, geology, and geotechnical engineering is a slippery slope, and engineering judgment will still be needed when dealing with Mother Nature and natural, non-uniform, and anisotropic materials.

Going back to governance: How representative do you think you are of the industry at large?

Everybody is going through a massive process with providing new guidance at industry, asset, and corporate levels. The Society for Mining, Metallurgy & Exploration (SME) had their first meeting about whether they should provide guidance. All this must funnel down to many accepted processes. The gold standard is to conform to the GISTM, incorporating lessons learned and promoting information sharing between companies.

Maybe we have arrived at the point where investors drive this; for example, they will not continue to invest in companies that do not adhere to leading practices. Even a small dam breach can kill people. There have been five tailings dam failures within one year after Brumadinho, yet no one is talking about them, even though one of them killed four people. It might fall to regulators and investors to not further these sites.

How do the tailings stewardship processes interface with other management processes such as health and safety and environmental?

For Anglo American, the 2018 annual sustainability report identified four catastrophic risks for the company, of which tailings dams is one. Underground fire and explosion, pit slope stability, and shaft integrity are the others. These all have special dedicated technical teams to manage and deal with the challenges posed. The tailings management team works with other teams for an integrated approach: social responsibility, environmental management, risk management, and health and safety teams. These integrations and interactions are internally reviewed for overall compliance and reporting.

A push and pull exists. The push comes from the technical center of excellence with specialists in, for example, closure, geochemistry, and geotechnical skills. Each of the experts collaborate with asset planning and design. The pull comes from assets that may have expansion plans, and peer review is needed, which requires people from safety, finance, and technical centers of excellence to work together.

Strong interfaces exist with various other management programs, especially safety and environmental, including sustainability reviews. The tailings culture is modeled after the safety personnel culture to mirror the high standards that "you take it home with you."

Outside stakeholders that are involved with noise and air quality related to the tailings facilities are not related to dam safety but get integrated into the tailings stewardship. Water is the oil of the 21st century. In the future, projects that do not save water, manage water responsibly, and meet our internal standards will simply not be approved.

How can we help the mining industry by spreading the message and sharing best practices?

It is getting better, and companies are more open to sharing and publishing. In the past, the stance was no news is better than good news. Most of the ICMM companies are providing lessons learned, and the hope is that it continues to improve.

Companies are preparing more papers, and more technical conferences are being offered. Mining companies are also sharing more disclosures with an increased amount of information being disseminated. Some mining companies are sharing best practices and lessons learned. Additionally, some mining companies engage in external research projects. However, it is always the same 10 or 20 mining companies. Where are the other 600 mining companies?

Our engineering firms are part of our training programs; we share lessons learned and we believe there is that flux of information. For example, a couple of our consultants and EoRs from South Africa visited a few Chilean facilities to understand more about processing sand for cyclone sand dams. We supported these efforts because this transfer of knowledge is helpful and important to us as a company. It is part of being leaders in the field. I believe that ICMM members are more open to these things. We do not want to have sharing barriers. That is part of the impetus to encourage the International Commission on Large Dams (ICOLD) to create and issue a bulletin—a guideline document that recognizes all dams around the world with an accepted standard of care that is not country specific.

Is there a forum for lessons learned sharing?

We need more of these focus forums. We also need to share our successes, for example, lessons learned about ways to monitor. In the past, companies have been reluctant to publish papers, but this has changed for the better in the last decade. To move forward, the industry appears to be more willing to share lessons learned—both good and challenging. It all takes time. It is up to all of us to get the message out about the successes and what we are doing well. The industry should be known for what we are doing well and not just the failures because there is a lot of good work being done. Additionally, it is time to charge ICOLD to update their 2001 bulletin on tailings dam failures.

CLOSING THOUGHTS

The tailings governance session outlined each company's approach to operational management and safety of tailings dams and water dams, with a focus on tailings stewardship. The tailings stewardship process or program is often closely tied to other management processes, such as health and safety, as well as environmental affairs and social license measures. Governance is generally managed from the corporate level and shared holistically with each site, with key input and auditing from third parties. Operating with a social license and engaging stakeholders to build confidence and earn their trust is instrumental. Recent tailings dams incidents, as well as several accidents and environmental events, have highlighted a need for diving deeper into governance processes.

The panelists emphasized the role for future information sharing to identify lessons learned (both good and bad) and acknowledge what the industry is doing well so as to not solely focus on the failures. In general, companies have come a long way in the last few years with being more open about sharing information, publishing disclosures, and allowing site access to build trust and repair industry reputation among the surrounding communities, investors, and the public at large.

Tailings Dam Failures and Regulatory or Social Response: A Reflection of Lack of Trust?

Kimberly (Finke) Morrison
Newmont, Denver, Colorado, United States

INTRODUCTION

Failures and incidents are often the catalyst for change, and this is evident with tailings dam failures and the inherent increase in regulatory control. The public's trust of the mining industry continues to degrade because of tailings dam failures, and, inherently, the regulatory environment continues to evolve with a corresponding increase in regulatory controls. This paper provides an overview of notable historic tailings dam failures that resulted in a dramatic regulatory or social response, illustrating the trust–control relationship as it pertains to tailings management.

NOTABLE FAILURES AND RESPONSES

Forensic studies to evaluate the mechanisms leading to tailings dam failures are highly studied and published, particularly by geotechnical engineers. However, the repercussions of failures from a social and regulatory standpoint often go unstudied by those individuals and, in many cases, may be considered underpublished. This paper presents several key tailings dam failures that resulted in dramatic changes to the regulatory framework in the jurisdictions for which they occurred; and, in some cases, the impacts were much further reaching, effecting a global response.

El Cobre No. 1 Dam Failure, Chile, 1965

The failure of the El Cobre No. 1 tailings dam in Chile after the March 28, 1965, magnitude 7.4 La Ligua earthquake with an epicenter 40 km away resulted in more than 200 fatalities (Valenzuela 2016; Penman 2001). From 1930 to 1965, the dam was constructed to a height of 33 m using the upstream method, with failure attributed to liquefaction of the sand tailings (Valenzuela 2016; Penman 2001).

At the time of this failure, many tailings impoundments associated with copper mining in Chile were constructed using the upstream method (Penman 2001). Two forensic engineers observed 22 tailings dams affected by the La Ligua earthquake, including the El Cobre No. 1 tailings dam, 11 of which failed or were badly damaged, finding that almost all of the failed dams had similar characteristics (Dobry and Alvarez 1967).

The El Cobre failure marked a turning point in Chile with regulatory agencies and the mining industry reacting by eliminating the option of upstream dams in favor of other solutions such as downstream-constructed dams despite the higher associated costs (Valenzuela 2016). Construction of tailings dams using the upstream method was prohibited in Chile as of 1970 through Decree 86 issued by the Ministry of Mining, unless a specific permit was issued by the national director of the agency for extraordinary conditions (Valenzuela 2016; Ministry of Mining 1970). However, it was not until 2007 that Decree 248 confirmed banning of tailings dams constructed using the upstream method in any instance (Valenzuela 2016; Ministry of Mining 2007).

Mufulira No. 3 Dam Failure, Zambia, 1970

Early on the morning of September 25, 1970, 2 million t (metric tons) of water and slimes from the overlying No. 3 tailings dam cascaded into the main working area of the Mufulira underground copper mine in the Zambian copper belt (Tembo 2019). Commencing in 1944, block caving methods were used to develop the mine, which led to development of a depression on the surface within which the No. 3 dam was contained, and, by 1956, this depression had subsided to form a basin with a maximum depth of 12 m (Tembo 2019). Not understanding the risks, mine management decided to fill the depression with additional tailings in 1963 (Tembo 2019).

This failure, which remains the worst mine disaster in the history of Zambia, killed 89 miners and cut the nation's copper production by as much as 50% (Tembo 2019). A historian of the incident stated that the failure "became a blessing in disguise because its aftermath set new standards in mine safety" (Tembo 2019). The Zambian mining regulations of 1971 (amended in 1973) banned the undermining of tailings dams (Republic of Zambia 1971, 1973). Specifically, Regulation 602 (Republic of Zambia 2019) states that

> No holder or manager shall mine or permit mining on or under any land where the intersection on the surface of the line of break due to subsidence from such mining operations will lie within a horizontal distance of one hundred metres from any…tailings dam…or any other object or surface feature requiring protection.

Stava Tailings Failure, Italy, 1985

On July 19, 1985, the tailings facility failed at the Prestavèl fluorite mine above the village of Stava, near Tesero, in northern Italy (Fondazione Stava 1985, n.d.). The dam of the upper basin failed and collapsed onto the lower basin, which also failed, releasing approximately 180,000 m^3 of tailings that flowed 4.2 km downstream, killing 268 people and destroying 62 buildings (Fondazione Stava 1985, n.d.).

Legal liability was sought after the failure, with the first trial taking place in Trento, Italy, and concluding in less than three years (Fondazione Stava 1985, n.d.). Legal proceedings ended after another four levels of judgment, with the second sentence pronounced by the Italian supreme court and delivered on June 22, 1992 (Fondazione Stava 1985, n.d.). This verdict confirmed the sentences passed by the first trial, whereby 10 people were convicted of multiple manslaughter and culpable catastrophe including persons involved in constructing the upper basin of the tailings facility during the period of 1969 to 1985, as well as regulators who failed to perform checks on the dams (Fondazione Stava 1985, n.d.). In the end, the imprisonment sentences were remitted, and no one sentenced served time in prison (Fondazione Stava 1985, n.d.).

Kingston Plant Fly Ash Failure, United States, 2008

The Kingston fly ash slurry spill occurred early on December 22, 2008, when an ash dike ruptured a 34-ha (hectare) solid waste containment area at the Tennessee Valley Authority (TVA) Kingston plant in Roane County, Tennessee. The failure released more than 3.8 billion liters of coal fly ash slurry and left up to 1.5 m of water and mud over 100–160 ha of rural land (Patel 2016). Just weeks after the Kingston spill, in January 2009, TVA experienced a failure at their Widows Creek Fossil plant releasing approximately 38,000 liters of gypsum slurry (Patel 2016).

Shortly after TVA's Kingston failure and subsequent Widows Creek failure, in March 2009, the U.S. Environmental Protection Agency (EPA) ordered review of approximately 300 utility coal combustion waste sites and announced that they would develop new rules to ensure that failures like the Kingston spill are not repeated (Patel 2016). Previously, regulation of these facilities was left to the states. In June 2010, the EPA issued a proposed rule regulating coal combustion ash either as a nonhazardous waste subject to tougher management and disposal requirements or as a "special" hazardous waste with similar controls but still eligible for recycling and reuse in products such as portland cement (Rickman 2010). The EPA administrator signed the "Disposal of Coal Combustion Residuals from Electric Utilities" final rule on December 19, 2014, and it was published in the *Federal Register* on April 17, 2015 (EPA 2015). This rule finalized national regulations to provide a comprehensive set of requirements for the safe disposal of coal combustion residuals from coal-fired power plants (EPA 2015).

Mount Polley Tailings Failure, Canada, 2014

On August 4, 2014, failure of the perimeter embankment at Imperial Metals' Mount Polley gold and copper mine released an estimated 10 billion liters of water and 4.5 million m^3 of tailings into the environment (CBC News 2014). Considered the largest environmental disaster in modern Canadian history (Breach of tailings pond 2014), this failure resulted in significant regulatory changes in Canada and elsewhere.

Such changes included reissuance of the Mining Code by the British Columbia Ministry of Energy and Mines with increased stringency, proposed changes to Canadian Dam Association guidelines for dam safety, and an industry review of the engineer of record (EoR) principle (GBA 2018). Mount Polley had five different individuals named as EoRs, representing three separate consulting firms in the four-year period prior to the failure (GBA 2018). The State of Montana was the first to respond to the EoR issue, enacting metal mine regulations introducing EoR and independent review requirements for tailings dams in 2015 (GBA 2018).

Samarco Fundão Dam Failure, Brazil, 2015

On November 5, 2015, the Fundão tailings dam at the Samarco iron ore mine, a joint venture of Vale and BHP, in the state of Minas Gerais (Brazil) failed with catastrophic consequences (Morrison et al. 2018). This tailings failure was only one of many documented in Brazil since 1986, while it was not until 2010 that Brazil enacted any laws pertaining to dam safety (Morrison et al. 2018). The Samarco failure released an estimated 43 million m^3 of tailings (80% of the total contained volume), polluted 668 km of watercourses extending from the Doce River to the Atlantic Ocean, and resulted in 19 fatalities (Morrison et al. 2018). One publication stated that "the collapse of Fundão was the biggest environmental disaster of the world mining industry, both in terms of the volume of tailings dumped and the magnitude of the damage" (Carmo et al. 2017).

The failure of Fundão effected significant changes to the regulations and requirements surrounding tailings management in Brazil, including overhauling the emergency action plan requirements to encompass downstream sirens and drills; establishing requirements for periodic dam safety reviews; requiring real-time monitoring of installed dam instrumentation using continuous video monitoring; and other requirements (Morrison et al. 2018).

In addition to regulatory changes, legal actions were taken. Samarco was subject to extensive litigation and government sanctions, and charges of manslaughter and environmental damage were filed in 2016 against 21 executives, including Samarco's former CEO and representatives from Samarco's owners, Vale and BHP, and the board of directors (Timson 2019).

This failure, combined with the Mount Polley failure the prior year in Canada, resulted in global initiatives as well. One such initiative led by the International Council on Mining and Metals (ICMM) involved development of a position statement establishing their framework for governance of tailings facilities (ICMM 2016). The ICMM represents 27 of the world's leading mining and metals companies and 36 associations; the joint venture partners of the Samarco mine, BHP and Vale, are both ICMM member companies.

Córrego do Feijão Dam I Failure, Brazil, 2019

The failure of Dam I at the Córrego do Feijão iron ore mine near Brumadinho, Brazil, on January 25, 2019, resulted in 270 fatalities and raised the industry's radar with respect to so-called inactive tailings dams (Morrison and Gomide 2019). The owner, Vale, was also involved in the Samarco failure just about four years prior; both dams were constructed using the upstream method.

Mere days after the failure, new regulations were imposed in the state of Minas Gerais, while a couple of weeks later, more strict regulations were proposed for tailings management within the entirety of Brazil (Morrison and Gomide 2019). The final regulation, Resolution No. 13 published in August 2019 by the National Mining Agency (ANM), bans future use of the upstream construction method and requires "decharacterization" of all existing upstream-constructed tailings dams (ANM 2019). Decharacterization includes decommissioning, physical and chemical stabilization, hydrological and hydrogeological control, and monitoring for the period necessary to confirm effectiveness (ANM 2019). Additionally, Resolution No. 13 provided other requirements regardless of dam construction methodology, including the following:

- Prohibits maintaining buildings within the self-rescue zone (10 km downstream of the dam)
- Prohibits construction of other dams or storage of radioactive waste in the downstream area
- Requires automated instrumentation with full-time monitoring for dams with an associated potential damage characterization of high
- Establishes minimum factor of safety requirements
- Requires water reduction measures (ANM 2019)

The Public Ministry of the State of Minas Gerais filed criminal charges against Vale and the engineering company TÜV SÜD and 16 individuals connected to them regarding the dam collapse (Public Ministry of the State of Minas Gerais 2020). Among those cited were Vale's former CEO and the TÜV SÜD engineers who signed the dam's statement of stability report months before the failure.

Failure of Dam I had further reaching effects, rapidly shaping the way the industry, its investors, and its insurers view tailings dam safety. Of note, the ICMM, the United Nations Environment Programme, and the Principles for Responsible Investment announced on March 27, 2019, that they would co-convene an independent Global Tailings Review to establish an international standard for tailings facilities. The *Global Industry Standard on Tailings Management* (GISTM) is now an ICMM member company commitment.

The failure at Brumadinho also marked an initiative toward greater global transparency. The Church of England Pensions Board and the Council on Ethics for Swedish National Pension Funds sent an urgent letter to 727 listed extractives companies on April 5, 2019, requesting specific disclosures on every individual tailings facility under their control. This request was part of an urgent global engagement by investors in response to the Brumadinho tailings dam disaster.

TRUST-CONTROL RELATIONSHIP

Failures and incidents are often the catalyst for change, as noted by the regulatory and social responses to the failures discussed in this paper. Recent failures continue to degrade the public's trust of the mining industry and, inherently, the regulatory environment continues to evolve with a corresponding increase in regulatory controls. This is demonstrated by the trust–control relationship illustrated in Figure 1.

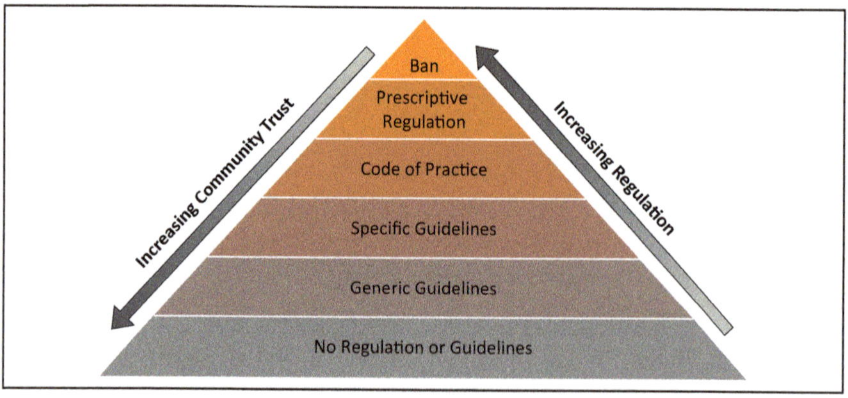

Source: Morrison and Gomide 2019

Figure 1 Trust–control regulatory relationship

REFERENCES

ANM (National Mining Agency). 2019. Resolution No. 13, August 8. *Official Gazette of the Federal Union*. Brazil.

Breach of tailings pond results in "largest environmental disaster in modern Canadian history." 2014. *Australian Mine Safety Journal.* August 12. www.amsj.com.au/breach-tailings-pond-results-largest-environmental-disaster-modern-canadian-history/. Accessed August 2020.

Carmo, F.F., Kamino, L.H.Y., Junior, R.T., et al. 2017. Fundão tailings dam failures: The environment tragedy of the largest technological disaster of Brazilian mining in global context. *Perspectives in Ecology and Conservation* 15:145–151.

CBC News. 2014. Mount Polley mine spill: Water quality test results within drinking guidelines. www.cbc.ca/news/canada/british-columbia/mount-polley-mine-spill-water-quality-test-results-within-drinking-guidelines-1.2730042. Accessed August 2020.

Dobry, R., and Alvarez, L. 1967. Seismic failures of Chilean tailings dams. Paper 5582. *Journal of Soil Mechanics and Foundations Division* 93(SM6):237–260.

EPA (U.S. Environmental Protection Agency). 2015. Disposal of coal combustion residuals from electric utilities. *Federal Register* 80 (April 17): 21301.

Fondazione Stava 1985 [Foundation Stava 1985]. n.d. 19th July 1985. www.stava1985.it/19th-july-1985/?lang=en. Accessed August 2020.

GBA (Geoprofessional Business Association). 2018. *Proposed Best Practices for the Engineer of Record (EoR) for Tailings Dams.* Rockville, MD: Geoprofessional Business Association.

ICMM (International Council on Mining and Metals). 2016. Tailings governance framework: Position statement. www.icmm.com/position-statements/tailings-governance. Accessed August 2020.

Ministry of Mining. 1970. Supreme Decree No. 86. Regulation of construction and operation of tailings dams. Chilean Ministry of Mining. August 13.

Ministry of Mining. 2007. Supreme Decree No. 248. Regulation for the approval of design, construction, operation and closure projects for tailings deposits. Chilean Ministry of Mining. April 11.

Morrison, K.F., and Gomide, P.C.R. 2019. Post-Córrego do Feijão: Continued evolution of tailings dam regulations in Brazil. In *Proceedings of Tailings and Mine Waste 2019.* Edited by J. Goodwill, D. van Zyl, and M. Davies. Vancouver: University of British Columbia. pp. 391–400.

Morrison, K.F., Pedrosa, H.G., dos Santos, G.J.I., et al. 2018. Changes to tailings dam regulations in Brazil in the aftermath of failures. In *Proceedings of Tailings and Mine Waste 2018.* Vancouver: University of British Columbia. pp. 27–34.

Patel, S. 2016. A brief history of U.S. coal ash since the Kingston spill. *Power Magazine* (March 1). www.powermag.com/a-brief-history-of-u-s-coal-ash-since-the-kingston-spill/. Accessed August 2020.

Penman, A.D.M. 2001. Risk analyses of tailings dam constructions. In *Seminar on Safe Tailings Dam Constructions*. Stockholm: Swedish Mining Association. pp. 37–53.

Public Ministry of the State of Minas Gerais. 2020. Criminal Investigative Procedure No. MPMG-0090.19.000013-4.

Republic of Zambia. 1971. Mining Regulations 1971, S.I. No. 107, May 21. *Government Gazette.*

Republic of Zambia. 1973. Mining (Amendment) Regulations 1973, S.I. No. 95, April 3. *Government Gazette.*

Republic of Zambia. 2019. *Guide to Mining Regulations: Consolidated Text of the Mining Regulations 1971* (S.I. 107/1971 and S.I. 95/1973). Lusaka, Zambia: Ministry of Mines and Mineral Development.

Rickman, J. 2010. EPA proposes two options to regulate coal ash. *Power Magazine* (June 1). www.powermag.com/epa-proposes-two-options-to-regulate-coal-ash/. Accessed August 2020.

Tembo, A. 2019. After the deluge: Appraising the 1970 Mufulira mine disaster in Zambia. *Historia* 64(2):109–131.

Timson, L. 2019. "Profit before people": Documents allege BHP execs were warned over deadly dam. *Sydney Morning Herald.* March 4. www.smh.com.au/business/companies/profit-before-people-documents-allege-bhp-execs-were-warned-over-deadly-dam-20190215-p50y6y.html. Accessed August 2020.

Valenzuela, L. 2016. Design, construction, operation and the effect of fines content and permeability on the seismic performance of tailings sand dams in Chile. *Obras y Proyectos* 19:6–22.

Tailings Governance and Management System: Freeport-McMoRan's Perspective

Teresa Speigl
Freeport-McMoRan Inc., Phoenix, Arizona, United States

Georgia Lysay
Freeport-McMoRan Inc., Phoenix, Arizona, United States

INTRODUCTION

Freeport-McMoRan Inc. (FMI) recognizes that the failure of tailings and other impoundments at any of our mining operations could cause severe, and in some cases catastrophic, property and environmental damage and loss of life. FMI's objective is to provide technical support and resources to have zero catastrophic structural failures of the 76 tailings storage facilities (TSFs) owned and operated by its subsidiaries. Many of our tailings impoundments are located in areas where a failure has the potential to impact individual dwellings, and a limited number of impoundments are in areas where a failure has the potential to impact nearby communities or mining infrastructure. As a result, our programs consider the significant consequences resulting from a potential failure, and we apply substantial financial resources and technical resources, both internal and external, to materially reduce the likelihood of failures.

FMI has developed and maintained a tailings management system (TMS) and stewardship program designed to be applied by our subsidiaries for continual improvement and assurance and to enhance the structural stability and safe operations of our TSFs. This paper discusses the evolution of the TMS and enhanced governance of the TSFs at FMI over the last 15 years. Specifically, four key programs from the TMS are discussed: tailings stewardship, technical review boards (TRBs), key performance indicator (KPI) reporting, and essential functions evaluation.

These programs are key to setting strategic direction, defining roles and responsibilities, as well as promoting alignment with corporate guidelines and procedures to foster ownership, transparency, and accountability among FMI operations.

In addition to these four key programs, the TMS also includes many other programs, processes, and document descriptions that provide consistency in operation and monitoring for our TSFs (e.g., annual reporting, operations and maintenance manuals, monitoring and action plans, standard operating procedures, and work instructions). The TMS allows for a cycle of planning, execution, review, and improvement while defining roles and encouraging collaboration between the site operations and engineering teams, corporate teams, and engineers of record (EoRs). The continuous improvement of the programs has allowed FMI to embrace the changing culture of TSF management in the broader industry and to proactively review and improve these programs and processes so they remain effective tools for our tailings operations and engineering teams.

TAILINGS STEWARDSHIP PROGRAM

In 2004, corporate executive leadership recognized the need for improved attention and understanding of our TSFs and launched the Tailings Stewardship Program (hereinafter the Program). This Program was the company's first formal corporate plan to provide support and resources for safe TSF operation at its subsidiaries. The original Program included development of a tailings stewardship team (TST), comprising internal and external engineering resources, whose focus was to conduct field inspections of all FMI TSFs on a periodic basis. Though the TST process has been refined over time, the TST has been consistent in their overall approach since the Program began. At each visit, the TST physically inspects all site tailings facilities (walking every toe, crest, and abutment); conducts a detailed review of operational, maintenance, and monitoring data; and meets with the site operations team. The TST then identifies potentially significant deficiencies and recommends corrective actions.

From its inception, leadership fully embraced the Program, and this was instrumental in the advancement of best practices. The TST was given substantial latitude to determine how to develop the Program; thus, the engineers worked with corporate leaders and site operators to develop fit-for-purpose procedures and documentation around stewardship functions. FMI leadership's strong commitment to the success of the Program, indicated by its provision of resources for implementation of actions to address findings and recommendations from the inspections, was critical to the acceptance by all levels of the operations—including frontline operators, supervisors, engineers, and mine site management. The excitement and passion for identifying and addressing challenges was contagious.

The acceptance of and commitment to the Program is evident in the transition of the content of the TST recommendations. Early visits resulted in numerous recommendations that required mitigation activities for potentially concerning conditions such as inadequate beach width or decants that had not been decommissioned. The Program drove enhancements to our EoR and site engineer inspections such that potential concerns now are proactively identified before they can become more severe. As a result, most TST recommendations now reflect conditions of a lower priority that can be addressed through ongoing maintenance programs.

The Program has grown significantly over the past 15 years and now also encompasses the reviews and inspections of FMI's 10 crushed leach pads and more than 300 water dam structures. In addition to collaboration with the EoRs, site engineers, and operations, the stewardship teams are supported by multiple geotechnical consulting engineering companies who are acting independently (not inspecting structures at which they serve as the EoR). In 2004, the TST conducted TSF inspections at six sites per year. This has increased to 41 site visits planned for 2020 including tailings dams, crushed leach pads, and water dams. To date, the TST has performed more than 1,000 inspections; walked more than 3,000 miles; generated greater than 1,400 recommendations; and taken over 150,000 photos of tailings, crushed leach, and water dams. The Program also offers a unique opportunity to foster internal resource networking, cross training, and sharing of best practices by inviting engineers and operating personnel from other FMI mine sites to accompany the TST on these inspections.

TECHNICAL REVIEW BOARDS

FMI has commissioned TRBs at all active TSFs and several inactive TSFs in our portfolio. The first FMI TRB was formed to support our Cerro Verde TSFs in 2004, followed by formation of TRBs for the North American sites between 2013 and 2016. Members of our TRBs visit our active and select inactive sites at least once every four years and meet with our internal teams and EoRs periodically to review our follow-up activities to their previous recommendations. We track recommendations from inception to completion.

The TRBs typically comprise three to four experts who are internationally known for specific and relevant areas of specialization. The TRB members typically have more than 30 years of experience and advanced degrees in applicable disciplines such as geotechnical engineering, water resource engineering, and seismology. They provide independent opinions and guidance on the physical integrity, safety, and performance of our TSFs and associated management systems. The TRB's terms of reference include identifying potential significant deficiencies with focus on geotechnical stability and water management. Whereas our TST focus is more on field inspections, operational procedures, organizational structures, training, and communications, the TRB's focus is often more on the work performed by the EoR and reviewing that the TSF is constructed and operated to meet the intent of the design. The TRBs utilize their experience in TSF design, construction, and operation; understanding of worldwide failure root causes; and knowledge gained from their professional activities to provide guidance and recommendations to FMI.

As with the TST, FMI leadership has embraced the TRB program. Each TRB for operating sites has direct access to the chief operating officer and direct access to the vice president of environmental services and sustainable development for inactive/legacy sites. FMI sites and EoRs address the TRB recommendations with the same priority and commitment as TST recommendations and have regular update meetings to review progress and action plans with the TRB.

Based on the recommendations provided by the TRBs, the sites and EoRs have conducted additional investigations to enhance material and site characterization. We also have improved our methodologies for engineering analysis to advance our understanding of the structural stability and performance of our TSFs.

KEY PERFORMANCE INDICATOR TRACKING AND DASHBOARD

One of FMI's most effective risk management and governance tools to support subsidiaries is our KPI tracking and early warning dashboard reporting, which was initiated in 2017. This tool is used for early identification of conditions that, if left unaddressed, may eventually lead to dam safety risks. The tool promotes proactive communication with leadership and provides assurance that routine operational construction, instrumentation monitoring, and maintenance activities meet the design intent and good practices for dam safety. The KPIs are organized into four categories:

1. Automated stability monitoring,
2. Reclaim pond management,
3. Dam construction/operation, and
4. Third-party action item tracking.

Examples of KPIs for each TSF include pore pressure levels, embankment movement, beach width, freeboard, embankment material quality, as well as internal and external inspections and implementation of recommendations received from multilevel review processes.

Scoring criteria for each KPI were established in consultation with the EoR and are based on many factors, including stability analyses, engineering judgment, and FMI guidelines. The criteria are based on confirming good practice and meeting the design intent of the TSF and so are reached well before an actual stability concern or emergency condition is triggered. The KPI scores feed into a dashboard report, which is a straightforward one-page summary of the four KPI categories. The mine site engineers compile the KPI data, review the dashboard report with the EoR, and submit the dashboard report monthly to their site's general manager and other management personnel, as well as key personnel in the corporate organization. In addition, this information is shared with FMI's corporate executive leadership each quarter.

The KPI process does not replace regular verbal updates or notifications between site engineers and management; rather, this process ensures a minimum level of open and clear communication.

This tool has been instrumental in promoting ownership, transparency, and accountability at the operational level. It also gives FMI's management, at both the mine site and corporate levels, the opportunity to review the condition of the TSF in a very simple and focused manner. This helps them to understand clearly where improvements are necessary for the proper management of their TSFs while still allowing detailed review of specific KPIs and supporting data when deemed appropriate.

ESSENTIAL FUNCTION EVALUATION

The essential function evaluation (EFE) program, initiated in 2019 at our active and select inactive TSFs, is one of FMI's most recent efforts to further improve our TMS. This program adds another layer of review and assessment, supplementing the governance programs and tools that already have been established to reduce risks associated with the TSFs. The EFE was developed to provide detailed guidance to site personnel and help management assess that day-to-day key functions of tailings management are completed with efficacy. The focus of this review program is primarily on the tasks of the site engineer, but it also considers tasks often performed by personnel in tailings operations, environmental, water resources, and site management. Examples of key functions include long-, medium-, and short-range planning, management and review of the EoR deliverables, field monitoring, material balance, water management, data and document management, inspections, and personnel succession planning and training.

The EFE is performed by a third-party reviewer familiar with FMI processes and expectations. The evaluation process consists of a review of documents related to mine site engineer activities; on-site interviews over multiple days with the mine site engineers, site tailings operators, and other departments supporting tailings-related activities; a closeout meeting; and a report documenting the findings from the review of EFEs at the site. The findings are shared with site and corporate management as well as the TST. Site personnel use the findings to identify priorities for continuous improvements, and the findings can be used as a tool for discussions between site engineers and their management on personnel and equipment resource allocation, support needs from other work groups, training needs, and organizational structure changes.

CONCLUSION

FMI's objective is for its subsidiaries to have zero catastrophic structural failures of our TSFs and, as such, over the last 15 years we have developed and maintained a TMS that incorporates many programs and processes, including those discussed in this paper. The TMS also defines the roles and responsibilities of all parties internal and external to FMI who are critical to executing these programs and processes and includes a range of personnel from site operators and engineers through corporate executive leadership. The TMS has been designed for continual improvement and assurance and to enhance the structural stability and safe operations of our TSFs. FMI has embraced the changing culture of TSF management in the broader industry and proactively reviews and improves all programs and processes, so that they continue to serve as effective tools for our tailings operations and engineering teams.

Tailings Dam Safety Management and Governance Framework at Anglo American

Caius Priscu
Anglo American, Vancouver, British Columbia, Canada

INTRODUCTION

The criticality of effective stewardship of tailings storage facilities (TSFs) has recently gained increasing recognition and triggered much effort on the part of the many mining companies across the world. The mining industry at global scale, and in particular the mine waste management practice area, is changing dramatically after the recent tragic events following the tailings dam failures in Canada (2014) and, more recently, in Brazil (2015 and 2019). Investors, communities, shareholders, nongovernmental organizations, regulators, and insurance companies require commitment to zero harm through the principle of no repeats, transparency, and implementation of best available practices (BAPs) and best available technologies (BATs). Anglo American is at the forefront of implementing solid governance programs, BATs, and BAPs across a large portfolio of TSFs.

THE ANGLO AMERICAN APPROACH

Anglo American's path to building a world-class mineral residue facilities (MRFs) practice includes many key aspects that are integral to a successful risk-informed approach:

- A solid and strict governance framework has been implemented for the oversight of a large portfolio of tailings and water dams in a global environment, where stringent minimum technical standards are implemented with non-negotiable minimum requirements.
- People with the right skills, competencies, and technical knowledge are needed, but also a safety mindset and a "do the right thing" approach needs to be part of their DNA.

- Critical controls are defined and implemented to ensure that each facility is planned, built, and managed safely to design requirements, minimizing the presence of the "surprise" effect and, in particular, the minimization of the human error factor.
- Continuous training of key personnel and learning from incidents (LFI) are core activities at Anglo American in this practice area, mandated internally through the technical requirements and acknowledged to be fundamental to maintaining competencies and advancing knowledge.
- Pushing the boundaries of technology and innovation in tailings production, deposition, and management of such facilities is necessary. From minimizing tailings production and water use to near real-time monitoring instrumentation and surveillance technologies, all are key to global success.

ANGLO AMERICAN TECHNICAL STANDARD

The new Anglo American journey in the area of dam safety and portfolio risk management of MRFs started in mid-2013, when the company completely revamped and transformed its internal technical requirements, raising the bar to a whole new level, aligned with leading world practices at that time. The technical requirements to allow rigorous risk management to take place at all MRFs within the group is specified in the Anglo American technical standard AA TS 602 001 (Anglo American 2019) and its associated technical specifications, which was rolled out in its new format in early 2014. The standard was updated as appropriate and goes beyond regulatory and other industry requirements in all host jurisdictions.

This mandatory internal global standard addresses the long-recognized principal risk that MRFs pose; sets minimum requirements for design criteria, monitoring, inspection, and surveillance; and was reviewed by international specialists.

The standard is applicable to all MRFs including tailings dams and water retaining dams that are either higher than 2.5 m or contain more than 25,000 m^3 of fluid and/or solids at maximum capacity.

Near the end of 2018, an implementation strategy was developed to provide full compliance to the standard, which constituted a five-year plan. In 2016, additional requirements were included in the same standard related to surface flooding risk management plans. This additional technical specification allows implementation of risk management programs to protect employees and contractors, operational assets, and mine working areas from floods and surface inrush.

When the International Council on Mining and Metals (ICMM) released its position statement on the governance of tailings facilities in December 2016 (ICMM 2016), the Anglo American technical standard was already fully aligned and, in some areas, went beyond the ICMM requirements. The Anglo American technical standard, considered best in class by external specialists and peers, is based on five pillars, which are the key focus points assessed to determine the internal levels of conformance.

Pillar 1: Consequence Classification of Structure Rating

Every structure in the portfolio must have a consequence classification of structure (CCS) rating based on failure consequence categories. Design criteria, monitoring, and surveillance requirements are set based on this rating.

Pillar 2: Competent Person

Every facility rated as having a high or major CCS rating must have a competent person (CP) in charge; the required competencies are described in the standard.

Pillar 3: Engineer of Record

Every facility must have an external engineer of record (EoR) to support the owner's team throughout the life cycle of the facilities: planning, design, construction, operation, and closure. During the operational phase, the EoR works continuously with the CP and management staff members to ensure construction and operational adherence to design and technical specifications.

Pillar 4: Technical Review Panel

A technical review panel (TRP) consisting of external, senior engineers is mandated for systematic and ongoing independent review. The TRPs are fully independent groups of specialists, with a variety of backgrounds, knowledge, and experience relevant to the respective area or business unit.

Pillar 5: Twenty-Five Required Documents

Anglo American requires a list of 25 documents and studies to be submitted. This is the absolute minimum necessary, ranging from design criteria and water balances to risk assessment and emergency response and preparedness plans.

ADDITIONAL GOVERNANCE PROCESSES

As of June 2019, the Anglo American portfolio of dams at managed operations included 91 TSFs and 68 tailings dams. Of all the TSFs, almost two-thirds are located in the southern Africa region (South Africa and Zimbabwe).

In addition, Anglo American operates more than 200 small to mid-size water retaining dams, ponds, and water diversion structures.

The oversight of the standard implementation is the responsibility of the Mineral Residue Facilities and Water Management (MRF and WM) central function based in Group Projects, Studies and Infrastructure, which is part of the Technical and Sustainability corporate division. The corporate team communicates and maintains close ties with each business unit (BU) and respective champions. The MRF and WM corporate team provides continuous project support and operational advice to the BUs and operational teams in this process with expertise, leading practice documentation, training, and the necessary on- and off-site guidance.

All new projects or expansion capital projects go through a rigorous investment assurance review to confirm fit-for-purpose design, appropriate design criteria, and best practice implementation throughout the project, meeting minimum requirements specified in our standard, as well as the investment models in effect.

Training is undertaken annually at carefully planned workshops and site visits, with participation from global practice area members, and including specialist guest speakers, as well as select EoRs and independent reviewers.

LFIs—both internal and external—are part of the risk management processes in place, aligned with the zero harm and zero repeats targets. Training is also carried out annually by the EoR on-site to ensure that personnel involved are familiar with the design intents, as well as the operations, maintenance, and surveillance manual.

CONCLUSION

Anglo American expects minimum adjustments to be included in its internal technical standard now that the *Global Industry Standard on Tailings Management* has been released (GTR 2020). In addition, an update of the internal water management standard is planned to take place in parallel to ensure compatibility and internal alignment of the two and their respective requirements.

The TSF risk management program at Anglo American is therefore a continuous cyclical process where structures are classified, work is prioritized, risk-informed decisions are taken, and mitigation measures are implemented as quickly and as effectively as possible.

The challenges in a short to mid-term range remain related to the availability of quality professionals in the field of tailings dam safety and management, and geotechnical engineering professionals ready to take the key roles at the operations around the world. Hence, training and capacity building continue to be a top priority at Anglo American, which ensures that a quality risk management and governance program for both tailings and water retaining dams is implemented at all levels within the company.

REFERENCES

Anglo American. 2019. Anglo American Technical Standard AA TS 602 001 Mineral Residue Facilities and Water Management Structures Standard and Technical Specifications. Original version released February 2014. www.angloamerican.com. Accessed August 2020.

GTR (Global Tailings Review). 2020. *Global Industry Standard on Tailings Management.* https://globaltailingsreview.org/wp-content/uploads/2020/08/global-industry-standard_EN.pdf. Accessed September 2020.

ICMM (International Council on Mining and Metals). 2016. Tailings governance framework: Position statement. www.icmm.com/position-statements/tailings-governance. Accessed August 2020.

Tailings Dam Safety and Governance at BHP

D. Chad LePoudre
BHP, Calgary, Alberta, Canada

Matt Currie
BHP, Melbourne, Victoria, Australia

Andy Haile
BHP, Perth, Western Australia, Australia

INTRODUCTION

Tailings dams are dynamic structures, and maintaining their integrity requires consideration of a range of factors, including appropriate engineering design, quality construction, ongoing operating discipline, and effective governance processes. Nothing is more important than the safety of our people and communities.

In total, BHP has 67 tailings facilities at our operated assets, 29 of which are of upstream design. Of the 67 operated facilities, we have five classified as extreme and a further 16 classified as very high. Thirteen of our operated facilities are active. The substantial inactive portfolio (54) at our operated assets is largely because of the number of historic tailings facilities associated with our North American legacy assets portfolio. At our non-operated joint ventures, we have nine tailings facilities. All non-operated facilities are located in the Americas.

Following the collapse of the Brumadinho dam at Vale's iron ore operation in Brazil in January 2019, BHP established an internal tailings task force team reporting to the executive leadership team and the board's sustainability committee. The task force is accountable for the continued improvement of our operated tailings storage facilities, progressing our technology efforts with respect to tailings management storage and participation, along with industry partners, in the setting of new international tailings management standards. In addition, we support calls for greater transparency in tailings management and plan to work with community, regulatory, and financial stakeholders to promote the application of consistent disclosure that informs better tailings dam stewardship.

DAM RISK MANAGEMENT

BHP assesses the dam classification, risk, and operational circumstances in determining whether to empower a tailings review or stewardship board (not all facilities have tailings reviews or stewardship boards). Tailings reviews or stewardship boards are either in place or in the process of being established for our operated assets with very high and extreme classified tailings facilities. BHP's approach to dam risk management at our operated dams is integrated into our standard approach to risk management, assurance, and continuous improvement with particular focus on four key areas:

1. Maintenance of dam integrity
2. Governance of dam facilities
3. Monitoring, surveillance, and review
4. Emergency preparedness and response

Supporting this approach to dam risk management at our operated assets are group-wide processes of technical support and oversight.

MAINTENANCE OF DAM INTEGRITY

Central to our approach is the recognition that maintaining dam integrity is a process of continuous assessment that needs to be maintained for the life (including into closure and post-closure) of a tailings facility. As a result, we have identified five key dimensions to maintaining dam integrity:

1. **Design.** The basis of dam design is guided by design criteria specified through the Australian National Committee on Large Dams, the Canadian Dam Association (CDA), and local regulation, as well as dam classification.
2. **Construction.** Quality assurance and quality control are maintained across all construction phases (from initial construction to dam lifts/expansions during operation to closure and post-closure).
3. **Operations and maintenance.** The dam is operated and maintained in accordance with its design requirements.
4. **Change management.** Any changes on dam design and integrity are identified, assessed, and mitigated.
5. **Monitoring, surveillance, and review.** We ensure that the dam is functioning as intended.

GOVERNANCE OF DAM FACILITIES

We believe that effective governance encompasses a range of aspects from the management of change in our business to appropriately employing and enabling qualified personnel with clear accountabilities. We have mandated three key roles across our operated assets, accountable to the asset general manager of the relevant asset. In addition, a central technical team (practice leads) enhances oversight and assurance.

- **Dam owner.** The single point of accountability for maintaining effective governance and integrity of the tailings storage facility throughout its life cycle is the dam owner. (A tailings storage facility can have multiple dams or cells that have a contiguous, structurally similar interconnected wall, operated under the same tailings disposal regime, and are interdependent for stability and of similar height and risk profile.)
- **Responsible dam engineer.** A suitably qualified BHP individual is appointed to be accountable for maintaining overall engineering stewardship of the facility, including planning, operation, surveillance, and maintenance.
- **Engineer of record.** An independent, suitably qualified professional engineer is retained by the dam owner for maintaining dam design, certifying dam integrity, and supporting the dam owner and the responsible dam engineer on any other matters of a technical nature.

MONITORING, SURVEILLANCE, AND REVIEW

Given that tailings dams are dynamic structures, we believe effective monitoring, surveillance, and review is central to ongoing dam integrity and governance. We believe these processes span six dimensions, with the level of utilization of each dimension depending on the specific needs of the relevant facility. These six dimensions include

1. Monitoring systems—operating in real time or periodically;
2. Routine surveillance—undertaken by operators;
3. Dam inspections—more detailed inspections undertaken periodically by the responsible dam engineer;
4. Dam safety inspections—annual inspections undertaken by the external engineer of record reviewing aspects across both dam integrity and governance;
5. Dam safety reviews—conducted by an external third party as set out in the following section; and

6. Tailings review or stewardship boards—a panel of qualified independent individuals, approved by and accountable to the asset general manager, whose capability is commensurate with dam significance, under specific terms of reference to review aspects such as the current status of the dam, any proposed design changes, and outcomes of any inspections or dam safety reviews.

The type and frequency of monitoring, surveillance, and review is informed by the consequence classification, complexity, and operational status of the dam. Dams that are likely to have a greater level of consequence because of failure, which have greater technical complexity and are actively operating, have monitoring, surveillance, and reviews with greater rigor and frequency.

DAM SAFETY REVIEWS

Dam safety reviews are central to our approach to dam integrity and continuous improvement. We engage an external engineer to undertake dam safety reviews consistent with the guidance provided by the CDA in their 2016 technical bulletin on dam safety reviews. As per this guidance, review frequency is informed by the dam classification under the CDA (2016).

Dam safety reviews are detailed processes that include a thorough review of dam integrity and dam governance and include a review of the dam break assessment and dam consequence classification. Reviews are led by an external qualified professional engineer (selected for the appropriate level of education, training, and experience), with support and input from other technical specialists from fields that may include, for example, hydrology, geochemistry, seismicity, and geotechnical and mechanical fields of study. At the conclusion of the review, the qualified professional engineer provides a signed assurance statement, which includes a comment as to the integrity of the facility.

EMERGENCY PREPAREDNESS AND RESPONSE

We believe the final key element in our approach to dam risk management is emergency preparedness and response. Our approach to emergency response planning for our tailings facilities is designed to be commensurate with risk, with the following steps taken as appropriate given the risk:

- Identifying and monitoring stability and operating conditions, with thresholds that prompt preventive or remedial action;
- Assessing and mapping the potential impacts from a hypothetical, significant failure, including infrastructure, communities, and environment, both on- and off-site, regardless of probability;

- Establishing procedures to assist operations personnel responding to emergency conditions at the dam; and
- Testing and training in emergency preparedness ranges from desktop exercises to full-scale simulations. Desktop and field drills are scheduled at a frequency commensurate with the level of risk of the facility.

REFERENCE

CDA (Canadian Dam Association). 2016. *Technical Bulletin: Dam Safety Reviews.* Toronto, ON: Canadian Dam Association.

SESSION 4

Tailings Continuum—Selection of the Right Tailings Technology

CHAIR

Rachel Jansen, Paterson & Cooke USA, Ltd.

PRESENTERS AND PANELISTS

Glen Barr, Twin Metals Minnesota
Jim Casey, Klohn Crippen Berger Ltd.
Davide Collini, Aqseptence Group S.r.l.
Christian Kujawa, NovaMetallix, Inc.
Todd Wisdom, FLSmidth, Inc.

NOTETAKERS

Kathy Adams, Paterson & Cooke USA, Ltd.
Sam Saunders, Freeport-McMoRan Inc.

INTRODUCTION

This session's presentations showed different perspectives and roles within the overall efforts to evaluate, select, and then implement the right tailings process technology across the tailings continuum. The panel discussion looked at present and near- and long-term projected trends within the continuum and provided all who attended an appreciation that "one size fits all" is not applicable to tailings deposition technologies.

PANEL DISCUSSION: TAILINGS CONTINUUM

Do you think larger filters are possible?

There is always room for improvement. A perception that filtered tailings are expensive exists because of two cost factors: maintenance and moisture removal. Technologies that allow blending of tailings with waste rock allow for less water removal. This process can place the fast filtering with GeoWaste™

composite cost at around US$1.50/t (metric ton), which is approximately half of the standard at US$3–$4/t. Cost savings can also be attributed to using conveyors versus trucks and so forth. Filtered tailings typically requires a long mine life to make it work. The key is not to consider only the cost of the technology, but also other associated costs and benefits.

What is the commercial availability of the large filter plates?

Five to six large filters can be produced per year with the current supply chain.

What upstream processes show the most promise for reducing tailings or producing better tailings? Are there any opportunities for returning tailings to saleable products?

At the Colorado School of Mines, by-products from tailings are being developed. For example, metals in tailings can be used as fertilizer. Metal reclamation can provide additional separation of gangue minerals. Legacy coal, iron ore, and phosphate tailings have value, but the right technology is needed. Upstream processes including ore sorting to reduce tailings volumes and separation of metals or tails components can be better utilized. The possibilities have not been fully explored: For example, why not utilize flat surfaces of reclaimed facilities as agriculturally viable land?

A Klohn Crippen Berger client in Alaska has barite in their base metal deposit, which may not initially be viewed as an economical commodity. Studies have identified that the barite may be economical for the project, and separation of the barite is being incorporated into their process flow. In addition to potential economic benefits, this can reduce the volume of tailings produced.

Additional things can be done through ore sorting: increase grind size (coarse-particle flotation) or eliminate or deal with fines from the regrind circuit separately from the main circuit. The coarser size improves dewatering of the tailings. Dry grinding should be considered, which entails sharper aggregate characteristics compared to that from semiautogenous grinding and ball mills, therefore reducing fines.

How can you deal with tailings variability using filter or tailings dewatering design?

Variability can come from different parts of the mine plan. The tail needs to wag the dog in this case. Good communication between the tailings facility and the ore processing and ore blending facilities is required to avoid impact to the back end. Good practice includes monitoring clay/fine production, grind control, and avoiding overgrinding.

Often, filtration plants are sized on a single sample. Variability testing needs to be completed prior to equipment selection. A conservative element to the design should cater to the outliers.

From an operations and design perspective, Greens Creek and Karara have redundancies built into their systems for operational upsets. For example, Greens Creek has a shed where they can divert tailings in bad weather. If a bad batch of tailings is received, Karara has a slurry facility in addition to the filtered tailings. They also have a conveyance system with a backup truck fleet delivery system. Layering in redundancies and resiliencies into the tailings operation is important.

At the design stage, good characterization of tailings and designing in redundancies is critical. Understanding what triggers reprocessing is imperative. Twin Metals Minnesota (TMM) has engineered in flexibility by building a flexible backfill schedule into the mine design. They also prioritize use of the material in the underground as backfill when necessary.

How do you see partnerships forming between vendors, operators, and consultants?

This is simply daily life, and the mindset we have. We should never consider our customers as customers; we should try to approach them as partners in developing solutions. We should ask for more samples, to better design for variability. We want the machine to work, and to meet the customer requirements. It is important to see customers as co-developers.

Partnership implies a risk reduction. Normally, the environment is extremely competitive between vendors; a bidding competition can prevent reaching the best solution. In a partnership, this risk is reduced, but partnerships require early engagement.

Partnerships have been key in the development of new technologies. Obtaining third-party reviews and securing university involvement can be a benefit. Partnerships are beneficial specifically in getting the first unit installed. No one wants to install the first or second unit of a new technology. It is important for us as an industry to collaborate more.

Regarding recent mining plans that combine two technologies, for example, filtered tailings and zero water discharge, how are these compatible? With recycling of process water, how do you prevent precipitation of salts in filter presses, and how do you keep the filter presses functional?

If the salts were to precipitate in the filter presses, this would be a blessing, as they are removed from the system. Usually, the salts stay in the reclaim water and go back to the process, causing issues with extraction. As mining operations move toward filtered tailings and zero discharge, most operations will require water treatment capabilities. Salts are generally beneficial for dewatering—improving thickener and filter operation—but usually cause a problem on the extraction side.

Keeping filter presses clean is standard, normal maintenance. Methods include cloth washing and flooding with water, or acidic/alkaline water, depending on the need. During system design, it is of utmost importance to know the expected chemistry.

Creating drier tailings for stability versus using a water cover on tailings to prevent acid generation are competing objectives. What role do you think technology plays in simultaneously addressing both these problems?

Sulfide segregation is a good opportunity for reducing the amount of tailings that can cause a geochemical issue on an unsaturated tailings beach. We need more research and development on tailings additives that promote crusting and oxygen ingress into dry beaches. Another opportunity is amendment, possibly marrying separate tailings streams that are complementary to resolve problems with either.

How does the TMM dry stack design address potential seepage into the groundwater?

Much like other operating facilities, the proposed design has underliner and overliner drains. The facility is sitting on competent and non-fractured bedrock. Once tailings are compacted, they have low permeability; appropriate water management supplements the drainage systems.

Process Considerations for Tailings Dewatering Across the Continuum

Christian Kujawa
NovaMetallix, Inc.

INTRODUCTION

Tailings dewatering is highly dependent on the nature of the tailings and the project-specific drivers that have a major impact on the final dewatering target moisture. The objective of this paper is to cover the pertinent aspects of tailings dewatering over the full dewatering continuum as required for the various tailings storage methods of conventional, thickened, paste, filtered, and engineered tailings. The emphasis here is on the fundamental role of the nature of tailings, thickening and mechanical dewatering technologies, and tailings storage requirements with the change in dewatering target moisture. Challenges that the tailings industry are currently facing are discussed, as well as the importance of new dewatering technologies and their potential.

THE TAILINGS PROCESSING OBJECTIVE

The ready availability of water during the inception of mining led to water becoming the vehicle of choice for comminution, hydraulic transport, and separation, a practice that is still common because of its cost effectiveness. Today, water is still the most abundant fluid on earth, but its use is contested by more stakeholders. Recovery of water from the tailings for reuse is becoming a focus, not only because of the value of water, but also because the fluidity that water can impart to the tailings makes tailings stored with too much water unsafe. The lower the final moisture content, generally the safer the final stored tailings are. The challenge of today is to achieve dewatering for huge tonnages, processing several thousands of tons per hour, and in the presence of either consistent or occasional clays, which do not dewater easily.

THE TAILINGS PROCESS

The main components that make up the tailings process are amendment, dewatering, transport, placement, and storage. The first three can follow any order. Depending on the final quantity of water associated with the tailings for placement and storage, each of these tailings process components can take on various forms and extent of implementation. The storage design of tailings falls squarely in the realm of geotechnical engineering.

Amendment can mean the destruction of unwanted chemicals (such as cyanide), the neutralization of acidic components, or the absorption of water (such as with cementitious binders). Dewatering is achieved using various driving forces such as gravitational (thickeners), vacuum, centrifugal, or pressure. Transport is either by pumping or the transport of filter cake by conveying or trucking. Storage can be as wet conventional tailings (for which several facility construction methods exist such as upstream, downstream, and central rise), paste tailings, or filtered stackable tailings.

THE DEWATERING CONTINUUM

Tailings are placed either as a wet slurry to drain (called *conventional tailings*), sitting by definition on one side of the dewatering continuum, or in a stackable form such as a paste or even as a filtered cake (considered a moist soil) (often erroneously labeled as "dry" tailings in the industry) on the other end of the dewatering continuum.

Given enough dewatering, a point is reached at which there is not enough water present to fill all the interstitial spaces between the particles. These voids are filled with air. It is at this point that the tailings move from a two-phase to three-phase arrangement per volume of tailings. The solids and water mixture at that point is defined as partially saturated. Several defined waypoints exist for this dewatering continuum in terms of delineated measurements as the water is progressively removed from the tailings, which are described next.

Wet Tailings

Wet tailings, whether thickened or not, are in slurry form and can generally be readily pumped.

Paste Consistency

Many definitions exist; however, the most pragmatic is the definition that paste has a fully sheared yield stress of 100 Pa or greater.

Zero Free Water

The zero free water (at 60 kPa vacuum), zero bleed water (around 16 kPa), or true filtration form step (from 100 to 1,500 kPa) filter cake moisture are similar. The difference in the measurement is mostly caused by the difference in driving force applied for the measurement.

Flow Moisture Point (and Transportable Moisture Limit)

This is the maximum moisture content beyond which the solid–liquid mixture liquefies on experiencing vibration. The flow moisture point is determined by an ASTM procedure with a flow moisture point table.

Standard Proctor

The standard Proctor is a geotechnical measurement. The ASTM measurement provides the maximum bulk density that can be achieved at an optimum compaction moisture content. The compaction effort is equivalent to a pressure of 600 kPa.

Modified Proctor

The modified Proctor is a similar measurement to the standard Proctor, but at a compaction effort equivalent to 2,700 kPa.

CHARACTERISTICS OF TAILINGS

Even tailings from a single mining industry tend to be different. Much like with rheology, depending on gangue mineralogy and liberation requirements, dewatering is much a function of particle size distribution, particle shape distribution, water chemistry, and solids surface chemistry, among others. For that reason, test work is an essential prerequisite to collect the needed information for design.

As every particle surface attracts an absorption water layer, the finer the particles, the greater the water associated with the solids. The finer the particle comminution, the larger the number of particles, but with an exponential growth of particle surface area. Clays especially tend to break apart into extremely small particles, often accumulating in the –2 µm size fraction. Unfortunately, clay particles are also often surface active, which imparts a level of chemistry interactivity, further complicating water removal.

Some mixtures, and especially those containing clays, are such that the filter cake produced tends to be compressible. Instead of making a rigid structure, the filter cake tends to compact under pressure. Compression of the tailings results in further dewatering, but most times also a reduction in the hydraulic conductivity of the tailings and subsequent rate of dewatering.

Nearly all dewatering test work is based on empirical confirmation and scale-up by direct correlation. There exists in filtration theory the almost forgotten parameter of specific cake resistance. The specific cake resistance, which must be measured as it cannot be computed from first principles, is nevertheless akin to a material property for any solid–liquid system. This immutable property can be used directly in scale-up calculation. In practice, the specific cake resistance range is from 10^8 to 10^{14} m/kg. Material with a specific cake resistance of 10^{12} m/kg and greater is considered difficult to filter. Recent work has shown that only technology capable of post-filtration steps such as mechanical squeezing and air blow drying can still dewater tailings with specific cake resistance of 10^{12} m/kg and greater.

Ore bodies are seldom homogeneous, and so life-of-mine variability throughput is likely. The challenge for the design of the tailings process is to cater for this variability.

TRANSPORT OF TAILINGS AND THE TRANSPORT DISCONTINUITY

Slurry tailings generally can be readily transported by pumping. The challenge is to keep the solids particles suspended and moving along in the pipe. As the tailings increase in solids concentration, the particles hinder each other because there is less and less space for particles to easily and smoothly maneuver without rubbing on other particles and the wall of the pipe. Thus, an increasing amount of yield stress is required to deform the slurry volume to move it. The yield stress of the slurry to resist deformation under application of a force increases as the solid concentration of the slurry increases, and the rheology is said to increase. There comes a point at which pumping of the slurry becomes so difficult that hydraulic transport of the tailings by pumping is no longer doable.

On the other end of the dewatering spectrum, filter cake that is low enough in moisture can be transported either by trucks or conveyors. The challenge here is to have a moisture content low enough that the tailings do not fluidize or liquify in the presence of vibrations. A filter cake that is too moist or wet cannot be transported by truck or conveyor without incurring risk of liquefaction. There exists, therefore, a region in which the moisture concentration is such that pumping is either too expensive or not feasible, but the mixture is too wet for transport by conveyor or truck. This transport discontinuity is therefore a problematic region from the tailings handling perspective and must be avoided.

DEWATERING TECHNOLOGIES

Although ubiquitous and inexpensive, gravity can be used to separate the liquid from the solids (as in thickening). Ever-increasing differential force is required to improve the separation, ranging from vacuum to centrifugation to pressure filtration (hyperbaric or direct hydraulic pressures), and ultimately drying (involving atmospheric, microwave, or even thermal drying).

Post-filtration technologies, such as a mechanical squeezing through a membrane press and direct air injection to overcome the capillary forces in a filter cake, or even drying methods, are employed to lower the final moisture further to get to the desired degree of partial saturation.

A plethora of mechanical dewatering technologies exist. However, many are not suitable for the treatment of large tailings tonnages nor necessarily capable of achieving partial saturation moisture targets.

TARGET MOISTURE IMPORTANCE

Because of the value of water to modern society and because of the role of water in the stability of tailings facilities, target moisture is the single most critical design parameter. It is the pivotal link between process engineering for dewatering and geotechnical engineering for the storage of the tailings. The moisture value has a major impact on capital and operating costs.

Geotechnical Criteria

From the geotechnical perspective, the lower the degree of saturation and the more compacted on placement, the less likely the stored mass will react to shock or vibration. Thus excess pore pressure in absorbed, and even in the event of a failure, risk is reduced because the stored mass moves less.

In the case of stacking of filtered tailings, many geotechnical companies have adopted the standard Proctor optimum compaction moisture as the proxy for the required final filter cake moisture target. The standard Proctor is, however, a challenging dewatering target to achieve for solid–liquid mixtures with a high specific cake resistance. The standard Proctor might be a useful communication tool. Ultimately, however, the specification of the geotechnically set target moisture should be based on more fundamental geotechnical principles.

Dewatering Cost Criteria

The effort required to recover increasingly more water from a solid–liquid mixture increases in order of magnitude as dewatering moves from gravity to mechanical dewatering to actual drying. Modification of the tailings (e.g., a coarser grind) such that a more relaxed (and more easily achievable) final moisture target applies significantly reduces the capital and operating costs. This is also true for the elimination of unnecessarily stringent moisture targets.

ENGINEERED TAILINGS

Engineered tailings as a concept encompasses all means of finding solutions to the prevention, reduction, or modification of the tailings with the express purpose of improving the processing of the tailings. At the most fundamental level, engineering of the tailings can entail ore sorting to reduce the amount of material exposed to water, coarser liberation to improve the dewatering characteristics, or converting the tailings to a by-product. Other refinements are possible in the defined tailings process, such as the following.

Split Tailings

The split tailings concept follows the divide and conquer principle. Most examples are of tailings being classified to produce two tailings streams for separate dewatering—a coarser particle size distribution, typically the larger stream, and a smaller stream of finer particle size distribution. The split is used to produce the coarse tailings that can be treated reliably with the more difficult dewatering curtailed to a smaller stream mass. Specific dewatering technologies are used for each stream. The two streams might or might not be combined again for final dewatering. No final lower moisture is necessarily achieved; in fact, most times, the final moisture is slightly higher. Invariably, however, the split treatment results in a much-reduced capital and operating cost.

Amended Tailings

In the simplest form, amendment of the tailings could simply be the addition of another but drier waste or tailings stream to produce a combined tailings stream that conforms with the set moisture target. A classic example of amended tailings is the addition of a cementitious binder to paste for underground backfilling. Amendment can take the form of eliminating unwanted ions from the tailings, such as in cyanide destruction. The addition of alkaline minerals to a tailings stream to render the tailings non-acid-generating is another form of amendment.

Goldcorp's/FLSmidth's EcoTails® tailings process solution falls into the amended tailings solution category. Dewatered tailings are co-mingled in the right ratio with a dry waste rock stream before deposition. The spaces in the waste rock are filled with the finer tailings, thus eliminating air and water flows that produce acid mine drainage, while the waste rock with the point-to-point contact between the large rocks provides inherent structure and stability to the stored material. Because the waste rock is dry, the dewatering of the finer tailings can be less intense, thus saving dewatering capital and operating costs. Disadvantages of this elegant solution are the rationing requirement of fine tailings to the waste rock and often the availability and logistics involved in achieving the co-mingling.

CLOSING THE LOOP—THE HOLISTIC SOLUTION

In the pursuit of improving water recovery for reuse and stable storage of tailings, the need for water treatment is introduced. Continued recycling of the water, with ongoing contact with fresh ore for processing, results in the steady increase in dissolved solids in the recycled water. Ultimately, the ion concentrations in the water impact the efficiency of the recovery of pay minerals. A level of water treatment is required to ensure that the quality of the process water remains acceptable to the process.

DEWATERING TECHNOLOGIES IN DEVELOPMENT

Traditional filtration has been found to be limited for solid–liquid mixture with a specific cake resistance greater than 10^{12} m/kg, where so far, only post-filtration technologies with high driving forces have shown some effectiveness. Possibilities exist in post-filtration. One such technology that could be integrated into filtration equipment is electro-osmosis. Alternatively, other technologies configured as an additional processing step, such as, for example, microwave drying with its focused heating of water, could serve as cost-effective post-filtration solutions.

Yet untapped are chemistry solutions other than the well-established fields of coagulation (chemical and polymer) and polymer flocculation. The introduction of hydrophobicity of solid surfaces could lead to much improved dewatering rates and likely stored tailings of higher hydraulic conductivities and therefore also consolidation rates.

CONCLUSION

Regardless of implementation, tailings process solutions need to be robust and reliable. To date, this has only been achieved with wet conventional, thickened tailings or paste backfill installations, which are relatively well understood. Filtered tailings have not found acceptance in the face of the demands of high tonnages or tailings with high specific cake resistance, worse even with the occasional incursion of clays in the tailings. Pure filtration has limitations making the use of post-filtration techniques and technologies necessary to achieve filter moisture targets with difficult to dewater tailings. New developments in dewatering are expected in post-filtration and the application of advanced chemistries. Ultimately, the optimal solution is one that is robust and yet provides the lowest cost profile.

Tailings Facility Design and Advancements in Tailings Technologies

Kate Patterson
Klohn Crippen Berger Ltd., Vancouver, British Columbia, Canada

Mary-Jane Piggott
Klohn Crippen Berger Ltd., Vancouver, British Columbia, Canada

Jim Casey
Klohn Crippen Berger Ltd., Vancouver, British Columbia, Canada

Over the last five years, there has been an urgency within the mining industry to improve tailings dam safety. A particular focus for many has been to move away from conventional tailings facilities, which can hold large amounts of water and flowable tailings, to alternative tailings technologies and facility types. These alternatives are centered around dewatering tailings prior to deposition (e.g., thickened, paste, and filtered) and placement or deposition of tailings within facilities that retain less water and non-flowable tailings (e.g., thin lift placement in a cone/downslope configuration, "dry" stack/pile, co-disposal with waste rock).

In addition to improved safety, other contributing drivers (i.e., potential advantages) that are often considered when adopting dewatered tailings or co-disposal practices include minimizing water losses for negative water balance mines, minimizing facility footprints, minimizing seepage, minimizing geochemical source loading from waste rock–only piles, and maximizing the potential to construct tailings facilities that can be transitioned to a landform that is suitable for passive closure quicker. Although there are advantages to these alternatives, some of the barriers that need to be considered include cost, constraints/limitations/complexity of operations (potentially caused by site conditions, production scale, tailings properties, etc.), geochemical considerations (e.g., dewatering may be at odds with the approach of maintaining potentially acid-generating tailings saturated; however, co-disposal with waste rock can minimize geochemical loading from the overall mine site), site-wide water management (e.g., often mine sites use their tailings facilities for seasonal water attenuation), and precedents for operations (i.e., the hesitation to be the "first" to try something new or push the limits of what is current practice).

Because of the increased interest in alternative tailings technologies and facility types, there have been continuous advancements in the application of these technologies in precedent-setting conditions. For example, we are seeing these technologies used on sites with higher production rates and finer tailings, and in wetter or colder climates. However, it remains true that every project and tailings facility has a unique combination of site conditions, tailings properties, available resources, social and regulatory environment, and countless other factors that must be considered and weighted when selecting a tailings management strategy.

An illustration of the key factors that should be considered when developing an optimal tailings management strategy with the goal of reducing the overall risk profile for all phases of mine life (design, operations, and closure) is shown in Figure 1 and described further in Table 1. To aid in the selection process, approaches, such as a tailings alternatives assessment including a multiple accounts analysis (MAA) combined with risk assessments, can be used to objectively and quantitatively evaluate the various aspects (i.e., technical, environmental, and social).

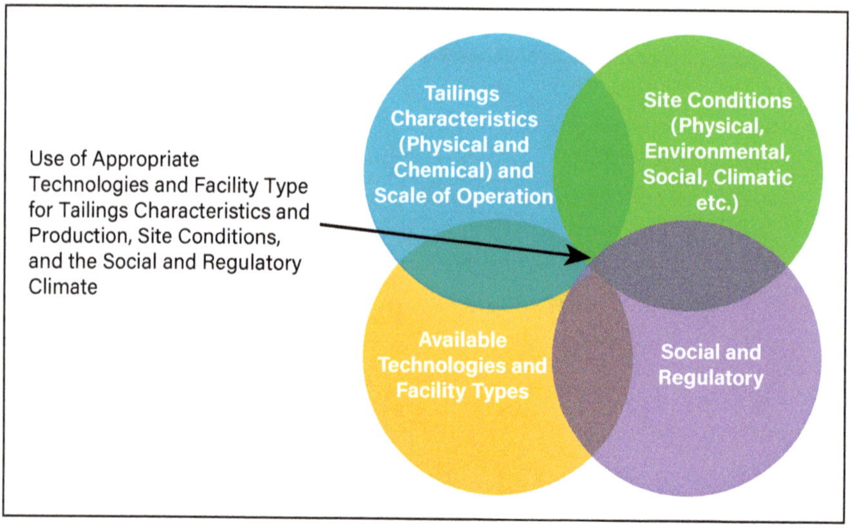

Source: KCB 2017

Figure 1 Tailings management strategy considerations

Table 1 Tailings management strategy considerations

Tailings Characteristics and Mine Plan

Tailings Physical Characteristics

- Finer tailings require more effort (power and time) to dewater to a higher solids content often required for transportation and deposition in co-disposal, thin lift, or filtered facilities.
- Tailings with high clay contents can be challenging to dewater (e.g., mature fine tailings [MFT]) and cannot be dewatered readily with commercially available equipment; the first full-scale centrifuge has been commissioned in the oil sands (Syncrude 2020) to attempt to address the challenges of dewatering MFT.
- Variability in ore/tailings needs to be considered in dewatering plant design targets (e.g., solids content/moisture content) and operations (e.g., Is backup storage required for off-spec material? Or is additional contingency required in processing capacity?).
- Tailings placed in filtered piles may need to be dewatered to low moisture contents for transport, placement, and potential compaction.

Tailings Geochemical Characteristics

- Leaching potential of tailings should be considered when deciding how to manage tailings. For example, tailings with sulfides are susceptible to oxidation, which may produce acidic drainage or neutral metal leaching. For these types of tailings, limiting impacts to the environment can be accomplished through inhibiting oxidation (managing the tailings when saturated, i.e., high moisture content and potentially submerged, limiting oxidation through compaction, covers, or encapsulation) or managing water that comes into contact with oxidized tailings (e.g., liners, seepage and runoff collection, water treatment).

Production Rate

- High-density thickening/thin lift placement in cone configuration and filter stacking have been successful at small to medium tonnage rates, but operations have encountered challenges when scaling up. For example, high-density thickening/thin lift placement in a cone configuration relies on achieving high solids contents in deep cone thickeners and often relies on achieving minimum beach slopes and large areas for drying tailings often to the point of desiccation. The challenges with scaling up this approach include
  - Scaling from operating one to two deep cone thickeners to many more, often result in inconsistent thickened product; and
  - Steep beach slopes can generally be achieved over a short distance, but tailings flatten over long distances resulting in larger areas required for the facility. Sites with footprint constraints may require containment dams, which limits the area and evaporative capacity.

Note: Several projects in arid environments have pushed the production of this approach to more than 100,000 stpd (short tons per day) to meet their mine plan and still meet the design intent. This indicates that given the right tailings properties, site conditions, and operations, the challenges of scaling up the production rate can be successfully managed.

- Another example includes filtering and stacking tailings. Challenges associated with filtered tailings, even at small scales, include processing to design targets (operating filter plants to produce consistent product for construction), transportation (trucks or conveyors, trafficability of filter piles, transport during poor weather conditions), erosion control (dust and water), and management of poor water quality. The challenges with scaling this approach include
  - Increasing the size and complexity of filter plants;
  - Increasing the transport distance for tailings and the active deposition areas;
  - With increased areas, increased management of erosion and water quality management is required; and
  - Interdependency of the operations.

Table 1 Tailings management strategy considerations (continued)

Note: One project in an arid environment in relatively flat terrain has pushed the production of this approach to more than 30,000 stpd to meet the design intent (Amoah 2019). This indicates that given the right tailings properties, site conditions, and operations, the challenges of scaling up the production rate can be achieved.

Water Balance and Site-Wide Water Management

- If projects are in a negative water balance, they are typically located in dry climates and a key objective is to minimize water losses. Conversely, if projects are in a positive water balance, they are typically located in wetter climates, and the main objective is to minimize contact water and safely manage water all over the site, also potentially requiring water treatment. Even in drier climates, projects can be in positive water balances, particularly seasonally, if the mine affected area is large.
- Mines often use tailings facilities as attenuation structures for the mine site's storm or seasonal water management, although this is generally seen as not best practice in modern tailings management. Adoption of dewatered or co-disposal tailings eliminates this potential and thus additional, or increased capacity of, water management facilities may be required. This means that when selecting a tailings management strategy, the mine site's water management strategy also needs to be reviewed.

Site Conditions

Construction Material Availability

- Quality and quantity of available borrow material is a consideration when selecting a tailings management strategy. Lack of quality dam construction material has been a driver for projects to select filtered tailings, particularly when the tailings themselves cannot be used as construction material in a conventional facility (e.g., cyclone sand).

Topography and Footprint Constraints

- Topography or terrain impacts the shape and size of a facility. Depending on tailings properties and scale, dewatered or co-disposal tailings strategies can be preferred or impractical. For example,
  - Filtered tailings may be preferred for small tonnage projects in rugged, steep, confined terrain with tailings that are not suitable for constructing a conventional facility; however, high tonnage rate projects typically transport the filtered tailings with conveyors to the tailings facility. These conveyors require gradual slopes and simple geometry, and therefore are not suitable for rugged, steep terrain.
  - Co-disposal, where co-mingling is conducted, also requires conveyors for mixing and transport, and thus is not well-suited for rugged, steep terrain.
  - As mentioned in the preceding "Tailings Characteristics and Mine Plan," high-density thickening/thin lift placement in a cone configuration typically requires large flat areas; this application is not conducive to rugged, confined terrain.

Climate

- Precipitation increases the moisture content of the tailings surface. Storm events can cause tailings surfaces to become not trafficable, which is an issue if the tailings surface needs to be accessed for deposition (e.g., filtered tailings) or if prolonged wet periods result in the design intent not being met (e.g., thickened thin lift deposition that relies on the entire tailings mass to have a low saturation percentage). However, small-tonnage filtered tailings projects in wet environments are managing trafficability by constructing roads from rock and with temporary contingency storage during wet periods.
- If a site has a high and reliable evaporation potential, there is potential to relax the target moisture content requirement and take advantage of the additional drying capacity of the site's climate.
- Drier tailings are more susceptible to dust. Small-tonnage dewatered tailings projects manage dust with compaction, temporary covers, wind fences, and so forth. Larger scale projects may have greater challenges managing dust because of the requirement to manage large active placement areas.
- Cold climates may present additional operational challenge during winter months for facilities that rely on mechanical equipment.

Table 1 Tailings management strategy considerations (continued)

Foundation

- Foundation conditions (e.g., geotechnical and hydrogeological) should be considered when deciding on a tailings management strategy. For example, if a site has a foundation that is not conducive to effectively controlling or capturing seepage, dewatered and co-disposal tailings facilities produce less seepage with the potential to enter the foundation and require management. However, dewatered and co-disposal tailings facilities may still require seepage management, even lining in some cases, particularly when the tailings produce seepage with poor water quality.

Environment (Water Usage, Groundwater, Waterways, Air Pathways, Habitat, etc.)

- No one technology or facility type is better for minimizing environmental impacts. Technologies and sites should be evaluated to determine which tailings management strategy minimizes environmental impacts.

Closure

- When constructing filtered tailings facilities, there is flexibility to create a preferred landform for post-closure.
- High-density thickened facilities that require large areas for active deposition cannot be progressively reclaimed and therefore require significant closure efforts over a large area at the end of operations.
- Draindown of the tailings is quicker with drier tailings.
- A strategy for managing sulfidic tailings can be to keep them saturated in the long term, which is not conducive to dewatered tailings.

Social and Regulatory

Consequence of Failure and Mobility of Tailings

- The primary concern of tailings management is the risk to public safety. Tailings facilities situated above populated areas have a high consequence of failure. The presence of a water pond and the mobility or flowability potential of the impounded tailings are factors in determining the extents of the potential runout of the tailings in a possible dam break. The tailings properties and water content determine the mobility potential or flowability potential of the tailings. If the tailings have a lower water content (e.g., dewatered) or are mixed with waste rock to change the combined properties, the mobility potential is much less, and the runout extent is therefore less. Regardless of technology or facility type, one should minimize the consequence of failure as much as possible, for example, through siting, configuration, reducing stored free water, and reducing water content within the tailings, and so forth.
- Recent research suggests that a GeoWaste* ratio of 1:1 for tailings-to-waste rock can be used for co-disposal to create a product that does not generate pore pressures if stacked to less than approximately 75 m (Bareither et al. 2018), although this can vary depending on tailings and waste rock properties at each site.

Regulatory Requirements and Guidelines

- Tailings facilities are regulated differently from country to country, state to state, and province to province; and regulations are often originally written for water dams or, in the case of liners, for landfills. Design and operating guidelines have also been established in multiple countries, states, provinces (e.g., Canadian Dam Association [CDA], Australian National Committee on Large Dams [ANCOLD], Mining Association of Canada [MAC]). Therefore, the regulations and guidelines used for tailings facilities are inherently variable. For example, in several jurisdictions, filtered tailings piles are not considered dams (particularly if they shed water) and thus may "slip through the cracks" of being regulated from a dam safety perspective.
- Although there may be regulatory advantages of certain technologies or facility types (e.g., facility is not required to be permitted through a dam safety regulator), international best practices should be used when designing, operating, and closing tailings facilities. This is especially true when considering the consequence of a tailings facility failure; usual practice is to use the methodology developed by the dam industry (even if the tailings facility is not considered a "dam").

Table 1 Tailings management strategy considerations (continued)

The Human Factor

- People can be the greatest asset but can also be a challenge for mining companies. The sentiments summarized in Erickson et al. (2017) succinctly describe the challenges associated with communications within an operating mine and with regulators. Communication challenges also extend to the public, consultants, corporate groups within mining companies, and so forth. Although these challenges are present with any tailings facility, they can be exacerbated with complex operations with narrow bounds for design targets and multiple opportunities for miscommunication. The philosophy of "Keep It Simple, Stupid" can help minimize communication challenges.
- Technologies and construction methods can vary in complexity, the level of operator skill required, as well as the number of skilled and/or unskilled workers required. The availability of this labor force for projects may be a significant consideration, especially for operations that are required to meet stringent, narrow-bounded design targets.
- Be it right or wrong, precedents for technologies and scale of operations are a consideration for mining companies. The public and investors demand high confidence in performance; therefore, there is a hesitation to be the first to try something new or push the limits of what is current practice.
- Complex technologies and construction methods require more controls and monitoring to ensure that design and construction targets are met.

Public and Regulatory Acceptance

- Regulations are intended to protect the public and environment. Technologies and tailings facilities that demonstrate public and environment protection should be acceptable to the public and regulators. Thus, evaluation of the preceding considerations and communication of those are important to gain that acceptance.

Available Technologies and Facility Types

Technologies Suitable for Tailings Physical Properties

- As discussed in the preceding "Tailings Characteristics and Mine Plan," not all technologies are suitable for all tailings types. Thus, selection of technologies or facility types is based on tailings properties.

Selection of Facility Type to Reduce Overall Risk of Tailings/Mining

- No one technology or facility type is best in all cases. The previously discussed considerations should be evaluated for each site to determine which tailings management strategy is best to reduce the overall risk of tailings management and the mining project.

*GeoWaste or Eco-Tails is a trademark term for a method of co-disposal being researched by a collaborative effort of Goldcorp, FLSmidth, and academia. GeoWaste or Eco-Tails is a blend of waste rock and fast-filtered tailings using filters and blending/transporting conveyors. Implementing GeoWaste or Eco-Tails is at a feasibility level in Canada and Mexico (i.e., proving concept at the operational scale).

This paper provides a high-level review of considerations for tailings facility design with an emphasis on advancing tailings technologies. This is not meant to be an exhaustive list of considerations, nor a road map for selecting a tailings management strategy. For a more detailed description of tailings technologies and the corresponding facility types, please refer to *Study of Tailings Management Technologies: MEND Report 2.50.1.* (KCB 2017).

COSTS

A key driver when selecting a tailings management strategy that is not included in the preceding is cost. Processing, transportation, and placement of dewatered and co-disposal tailings generally cost more than conventional tailings, such that implementation of these technologies may render some projects financially nonviable. For example, case histories indicate that mill-to-placement costs of filtered tailings can range from approximately US$2.5/st (short ton) to more than US$17/st.

However, standard cost estimates and tracking by mines do not always account for environmental impacts or potential consequence of tailings facility failure. They also generally grossly underestimate closure costs and post-closure maintenance costs; and often discount closure costs, rendering them a minor consideration. Thus, cost estimates alone cannot always be used as a representation of true value.

When selecting a tailings management strategy, one should consider the merits of the factors presented in Figure 1 and Table 1 in addition to cost. An MAA provides an approach that mining companies can use to weigh these values and aid in making a tailings strategy selection.

CONCLUSION

Continuing advancements in technologies are pushing the previous limits of tailings management and addressing some of the barriers to implementing alternatives to conventional tailings facilities. However, it remains true that every project and tailings facility has a unique combination of site conditions, tailings characteristics, available resources, social and regulatory environment, and countless other factors that must be considered when selecting a tailings management strategy, approaching design and management throughout the project life cycle. Selection should be based on evaluation of the full risk profile of alternative tailings management strategies (including life-cycle costs); there is no one-size-fits-all approach to tailings management.

REFERENCES

Amoah, N. 2019. Large-scale tailings filtration and dry stacking at Karara magnetite iron ore operation. In *Proceedings of Tailings and Mine Waste 2019*. Edited by J. Goodwill, D. van Zyl, and M. Davies. Vancouver: University of British Columbia.

Bareither, C.A., Gorakhki, M.H., Scalia, J., et al. 2018. Compression behavior of filtered tailings and waste rock mixtures. In *Proceedings of Tailings and Mine Waste 2018*. Vancouver: University of British Columbia.

Erickson, B., Butikofer, D., Marsh, A., et al. 2017. Filtered tailings disposal case history: Operation and design considerations, part I. In *Proceedings of Tailings and Mine Waste* 2017. Vancouver: University of British Columbia.

KCB (Klohn Crippen Berger). 2017. *Study of Tailings Management Technologies: MEND Report 2.50.1. Mine Environmental Neutral Drainage (MEND) Project*. http://mend-nedem.org/wp-content/uploads/2.50.1Tailings_Management_TechnologiesL.pdf. Accessed August 2020.

Syncrude. 2020. Tailings. www.syncrude.ca/sustainability/tailings/tailings-reclamation/centrifuged-tails/. Accessed October 2020.

The Future Is Now

D. Collini
Aqseptence Group S.r.l., Lugo, Italy

A. Pezzi
Aqseptence Group S.r.l., Lugo, Italy

D. Stenglein
Aqseptence Group Inc., New Brighton, Minnesota, United States

INTRODUCTION

Mining tailings have been and continue to be stored at most mine sites as a slurry. In this century-old process, the slurry is stored in a tailings pond, which is typically constructed on fabricated embankments. These embankments can be of various design, but nonetheless carry risk of failure. As recent events have shown, dam failures can lead to significant economic loss, environmental destruction, and, most importantly, loss of life.

Is there an option to tailings pond storage? Yes, and it is aptly called *dry stack storage*. The process has been in use since the 1910s. The *Global Tailings Standard* (hereinafter the Standard), which is the recent initiative raised by the Global Tailings Review (GTR), defines 17 principles that globally aim to achieve the safe and secure management of mine tailings facilities (GTR 2019, 2020). Dry stack storage, enabled by filter press technology to dewater the tailings up to solid cakes with extremely low residual moistures, perfectly fits Principle 5 of the Standard. This current best available technology is often dismissed because of cost. However, none of the recent dam failures would have occurred if the tailings were dry stacked. As the focus on tailings management changes, so does the technology.

FILTER PRESSES

Tailings dewatering has become a focal point today and experienced manufacturers have tailored their product line for this application. Currently, the Diemme Filtration GHT2500F is one of the filter press models available for large-scale dewatering projects, designed specifically for tailings dewatering

Courtesy of Aqseptence Group S.r.l.

Figure 1 Diemme Filtration GHT2500F filter press for tailings dewatering

(Figure 1). Since 2013, more than 50 GHT2500F units have been sold for tailings dewatering applications. Our largest multi-filter solution dewaters up to 25,000 t/d (metric tons per day) of tailings.

More and more often, the dry stacking operation using filter presses is considered a convenient tailings management solution when compared with the option of extending an existing tailings dam or even building one anew. Filter press technology offers suitable performance in terms of capacity and cake dryness. Economically, it is more competitive, especially when taking climate and other environmental factors into account.

Besides the environmental and safety issues that dry stacking tailings can solve, water is increasingly a key resource to manage, particularly in areas with frigid climate or freshwater scarcity (Figure 2). Filter press technology allows an efficient and early recovery to the process.

CASE STUDIES

Two case studies were performed using the filter press technology. The first case study, in the Chukotka region in Russia, used an auriferous mineral with tailings mainly consisting of quartz and orthoclase with a fine and regular particle size distribution [Dv(50) = 24 μm and Dv(90) = 101 μm]. To maximize the water recovery and avoid thixotropy in the cake, the filter press required a variable-volume plate pack. A final cake blowing step was needed to create a brittle cake that would be easy and safe to manage. Table 1 shows the key figures of the installation.

Courtesy of Aqseptence Group S.r.l.

Figure 2 GHT2500F filter press in gold tailings application (Chukotka region, Russia)

Table 1 First case study of a filter press plant: dewatering gold tailings

Number of filter presses installed	4 × GHT2500F20
Filtration area per unit	830 m^2
Cake volume per cycle	16 m^3
Plates (mixed-membrane plate pack)	91 (40-mm chamber thickness)
Duty (suspended solids [SS])	225 t/h
Feed SS concentration	45% w/w
Average cycle time	16 min
Cake specific gravity	2 kg/dm^3
Bulk density (after discharge)	1.5 kg/dm^3
Final cake moisture	15% w/w
Filter cake production	132 m^3/h

Courtesy of Aqseptence Group S.r.l.

In the second case study, in the Asia-Pacific region, a nickelous mineral was used. The tailings mainly consisted of hematite, gypsum, and goethite with a fine and regular particle size distribution [Dv(50) = 11 μm and Dv(90) = 130 μm].

An extensive sampling and testing program was developed for this specific case, which included laboratory thickening and filtration tests and a long site-based pilot-scale study.

To properly design the stacking area, it was decided to fast track the supply of one filter press, that has run successfully for three years producing 204,318 t (metric tons) of dry cake over 7,760 filtration cycles, with zero accidents. The remaining nine units are now under installation.

Even if the filter press cake moisture is in the perfect range to grant consistency and resilience to the stacking, the site receives more than 3 m of rainfall each year on average. Thus, the structural design of the stacking area must be carefully assessed to avoid liquefaction of the entire mass. Because of the long-running attention to stacking safety at this site, the filter cake from the new dewatering facility will be conveyed and stacked up to 100 m high.

The new tailings dewatering facility replaces the existing tailings dam, which was approaching capacity. The dry stacking system will allow safe operation of the refinery well into the future. Table 2 shows the installation and the key figures of the Asia-Pacific case study.

Table 2 Second case study of a filter press plant: dewatering nickel tailings

Number of filter presses installed	10 × GHT2500F22
Filtration area per unit	930 m^2
Cake volume per cycle	19 m^3
Plates (mixed-membrane plate pack)	101 (43-mm chamber thickness)
Duty (suspended solids [SS])	835 t/h
Feed SS concentration	35% w/w
Average cycle time	15 min
Cake specific gravity	2 kg/dm^3
Bulk density (after discharge)	1.5 kg/dm^3
Final cake moisture	28% w/w
Filter cake production	1,730 m^3/h

Courtesy of Aqseptence Group s.R.L.

FUTURE OF TAILINGS DEWATERING

The trend of tonnages for tailings, however, is on the rise because of lower head grades coupled with larger, more efficient mining and upstream processing equipment. As a result of this trend, the future of tailings dewatering

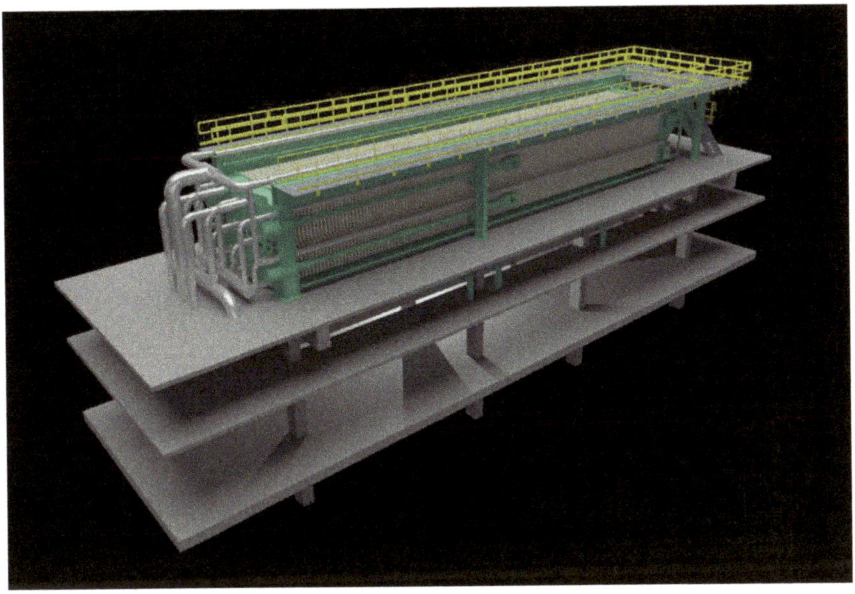

Courtesy of Aqseptence Group S.r.l.

Figure 3 Rendering of GHT5000F filter press

requires larger models. The Diemme Filtration GHT5000F responds to this need by upgrading the GHT-F family to capacities ranging from 12,000 to 15,000 t/d (Figure 3).

Table 3 compares the two filter press models: the older versus the newer. This step-change in filter press capacity presented many challenges to overcome. The most significant was associated with the filter plate design. Filter plates are typically manufactured of polymers and as such are molded. With limitations on the mold sizes available, our engineers considered a plate design differing from traditional designs. Further complicating the plate design was the transportation and handling of the plates. The result is a filter plate that incorporates segments of a plate to be fastened at the site.

Table 3 Filter press model comparison

Filter Press	Working Pressure Range	Maximum Cake Volume	Maximum Filtration Area	Maximum Length	Weight
GHT2500F	0-15 bar	26,800 dm^3	956 m^2	28 m	100,000 kg
	0-220 psi	10,145 gal	10,290 ft^2	92 ft	220,460 lb
GHT500 F	0-15 bar	70,000 dm^3	~2,900 m^2	37 m	400,000 kg
	0-220 psi	18,492 gal	30,460 ft^2	121 ft	881,850 lb

Courtesy of Aqseptence Group S.r.l.

Overcoming this physical hurdle, however, created additional hurdles associated with the fluid dynamics of the design. With the aid of computational fluid dynamics tools and modeling, our engineers developed a filter plate with innovative design, providing optimized fluid distribution inside the machine, improving the performance, and reducing the abrasion of all components in direct contact with the product.

The filter press structure was also a hurdle. As the filter press size increases, so does the amount of structural steel, thereby increasing the cost. To minimize the amount of steel, both finite elemental analysis and topology optimization were performed on the structure. The result is a strength-optimized structure saving approximately 7% of steel material.

CONCLUSION

In conclusion, the future of tailings management is trending toward dry stack methods. Thanks to the innovations of manufacturers, the cost and applicability of the equipment for this purpose is improving and is paving the way toward wide-scale implementation.

If lagoons, dams, and ponds are considered conventional tailings management solutions, then we think it is time to introduce a new terminology: "modern conventional tailings management." Using the new validated solution of filter presses and dry stacking, modern conventional tailings management grants an effective, safer, and more sustainable storage of tailings.

REFERENCE

GTR (Global Tailings Review). 2019. *Global Tailings Standard: Draft for Public Consultation.* https://globaltailingsreview.org/wp-content/uploads/2019/11/EN-Global-Tailings-Standard_CONSULTATION-DRAFT.pdf. Accessed November 2020.

GTR (Global Tailings Review). 2020. *Global Industry Standard on Tailings Management.* https://globaltailingsreview.org/wp-content/uploads/2020/08/global-industry-standard_EN.pdf. Accessed November 2020.

Recent Developments in Tailings Dewatering Technology

Todd Wisdom
FLSmidth, Inc., Salt Lake City, Utah, United States

INTRODUCTION

Tailings are commonplace in mining. Most tailings are stored within an impoundment contained by a tailings management facility (TMF). However, TMF failures are the most significant environmental liability for a mining project. With mining operations now pursuing lower grade deposits, higher throughputs are required for profitability. A safe and cost-effective solution for TMFs capable of accommodating 150,000 t/d (metric tons per day) are needed. On average, over the last 30 years, TMFs have experienced ±20 dam failures per decade, and a third of them caused serious safety and environmental liabilities, including deaths.

Several dam failures in Brazil in recent years have highlighted the consequence of TMF failures. In terms of the human, economic, and environmental impact and cost, with reparations and liabilities running into the billions of dollars, not to mention the reputational damage and market capitalization devaluation, disasters such as these major dam failures must be eliminated.

Access to water is becoming increasingly difficult. Mining operations have seen water costs steadily rise. Water shortage and water contamination are challenges miners have faced for some time, and this trend will only grow in the years ahead, creating huge financial and operational risks. About 70% of mines operated by the Big Five—BHP, Rio Tinto, Anglo American, Vale, and Glencore—are located in countries where water stress is considered a major risk.

PRESSURE FILTER CHALLENGES

Although dewatering tailings using pressure filters has been commonplace for some time, traditional filtered tailings costs are currently too high for ore deposits that are high throughput and low value per metric ton. Typical filtered tailings using smaller pressure filters and air drying the filter cake to an optimum cake moisture content, and then using trucks for tailings transport and placement has an in-place cost of more than US$4/t (metric ton). For large-scale, low-grade open pit mines, this cost is too high.

To reduce costs associated with dewatering tailings using pressure filter technology, three main problems need to be solved:

1. Low filter press volume per filter (filter presses are a batch process and low volumes per batch mean low production rates per filter)
2. Intensive maintenance and low availability (impacting operating costs and requiring high numbers of spare filters to maintain filtration plant availability with plant availability)
3. Slow cycle times (slow cycle times mean fewer batches per hour and lower production rates)

Low Filter Press Volume per Filter

The solution to low filter press volumes is to increase the plate size and increase the number of filter chambers in the filter press. Larger filter volumes have many benefits:

- Fewer filters
- Smaller building footprint
- Fewer support components, for example, pumps, valves, instruments, controls, piping
- Fewer cake-handling conveyors under filters
- Simpler control coordination between filters
- Fewer filter chambers
- Reduced number of filter cloths to change
- Reduced number of filter plates to handle

FLSmidth has designed a filter that includes a filter plate with the dimensions of 5-m high by 3-m wide and with 160 chambers, resulting in a filter volume of more than 90 m^3 and over 4,000 m^2 of filtration area. Figure 1 shows a rendering of the 5×3 plate, and Figure 2 shows a rendering of the 5×3 filter as currently designed.

Intensive Maintenance and Low Availability

Most maintenance on a filter press is associated with the filter cloth and filter plates. Filter cloth has a typical lifetime of around 3,000 cycles on most mineral flotation tailings. This leads to changing the filter cloth on the filter approximately twice per month, depending on how often the filter is operated. A plate design, shown in Figure 3, eliminates the typical failure areas on the filter cloth allowing much longer cloth life.

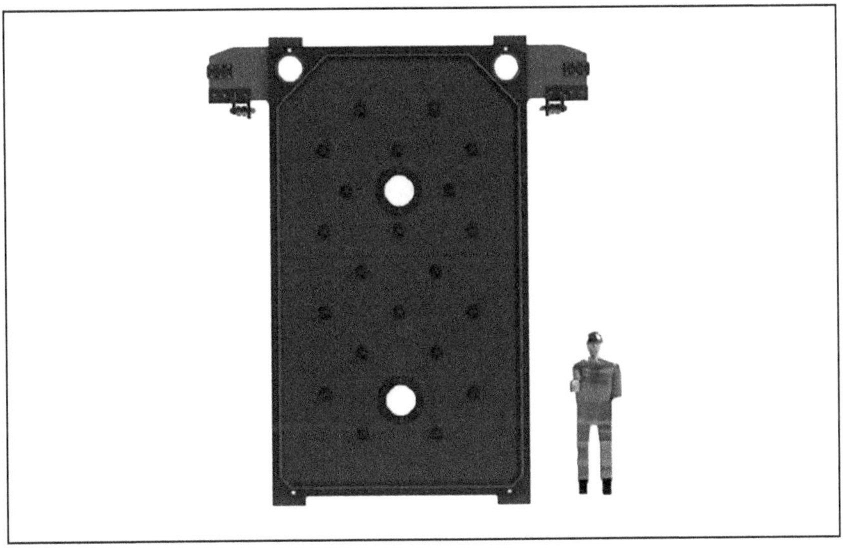

Courtesy of FLSmidth

Figure 1 Plate (5 m × 3 m)

Courtesy of FLSmidth

Figure 2 Filter (5 m × 3 m)

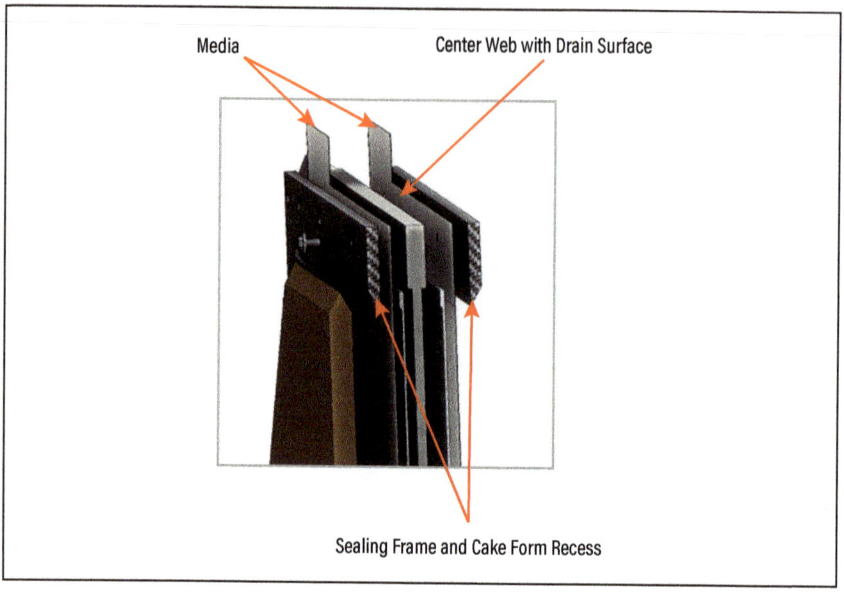

Courtesy of FLSmidth

Figure 3 Plate assembly (5 m × 3 m)

As was described in Wisdom (2019), high availability can be achieved by performing the cloth and plate maintenance outside the filter press. Current filter presses can achieve more than 90% availability utilizing this method. It is envisioned for the 5×3 filter that plate cassettes of 10 are removed, as shown in Figure 4, and maintained outside the filter. As soon as the dirty filter plates are put into the maintenance rack, new clean plates are lifted into the filter allowing the filter to operate while the dirty plates are maintained. As the cloth and plate maintenance is performed outside the filter, while the filter is operating, high availability of the filter is maintained. The total downtime to remove a set of plates and reinstall a set of plates has been measured in an operating filter to be 10 minutes.

Slow Cycle Times

Filter cycle times consist of two main components—process time and mechanical time—as described by Rahal (2019). Mechanical times are minimized by

- Opening the complete plate stack all at once,
- Maximizing the speed of the opening and closing,
- Performing valve openings and closings in a parallel method when possible, and
- Optimizing cloth washing and cloth shaking times when possible.

Courtesy of Ausenco

Figure 4 Plate cassette removal

Process times are minimized by

- Filling the filter at high pressures,
- Eliminating the use of membranes for squeezing the cake,
- Reducing or eliminating cake air blow, and
- Optimizing feed pumping control to take full advantage of pump flow and pressure capacities.

For most flotation tailings, total cycle times of less than 8 minutes can be achieved if both the process and mechanical times are minimized. The fast cycle times and a rapid filtering process is referred to as *fast filtering* and allows more batches per hour and higher unit filter capacities. However, if the cake air blow time is reduced, it may also reduce the volume of water removed from the filter cake produced. The resulting higher cake moisture content, typically between 18 and 20 wt % moisture, introduces potential geotechnical issues within the TMF.

EcoTails Process

To give the TMF the required strength needed when the higher cake moisture tailings are deposited, the EcoTails process co-mingles waste rock with the tailings. Mixing the low-moisture waste rock with the higher moisture tailings gives a product with an average moisture that is acceptable for stacking. The coarser rock particles also add shear strength and provide a higher density to the pile (Figure 5). This mixture of tailings and waste rock is called GeoWaste.

Historically, co-mingling has been challenging. Typically, 4–5 t of waste rock are required for each metric ton of tailings, and blending the resulting large volume of material has been challenging. Transporting large waste volumes is also unattractive to mining companies. With EcoTails, much less waste rock is required because the tailings are dewatered, and blending can be accomplished during conveying transportation. EcoTails mix ratios are designed for individual mine site requirements.

Another benefit of mixing tailings with waste rock is to reduce the void ratio and permeability of the waste rock pile. The lower void ratio coupled with a near-optimum moisture content has the potential for reducing acid rock drainage (ARD) within the TMF. ARD requires water and oxygen to thrive. Waste rock piles with the potential to generate ARD typically experience acid drainage because seepage is exposed to oxygen. Adding saturated filter cake tailings to the waste rock pile reduces the waste rock permeability. Oxygen flow through the pile is prevented and ARD can be greatly reduced.

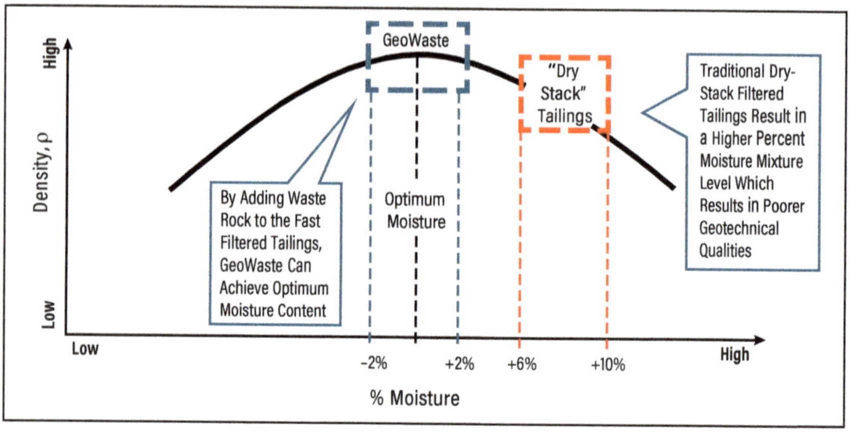

Courtesy of FLSmidth

Figure 5 GeoWaste optimum moisture

The EcoTails process, which produces GeoWaste designed specifically for each site, can produce a material that is geotechnically and geochemically stable.

GeoWaste Product

For EcoTails to work economically, GeoWaste must be produced in a continuous manner. Traditional tailings and waste rock co-mingling has been around for more than a century but used trucks and shovels to mix the two streams together. This is too expensive for a large tonnage open pit mine. The concept for GeoWaste is to use conveyors for transport and use the energy from the conveyors and specially designed transfer chutes to mix the material. Figure 6 depicts how GeoWaste can be implemented.

In 2018, a study comparing co-mingling using trucks and shovels versus conveyors was completed and indicated that the conveyor method is superior (Wisdom 2018). Figure 7 shows GeoWaste produced using conveyors and transfer chutes for mixing.

CONCLUSION

Processing more ore to recover needed metals as mineral reserves continue to decline directly increases the amount of tailings produced from mining. Water

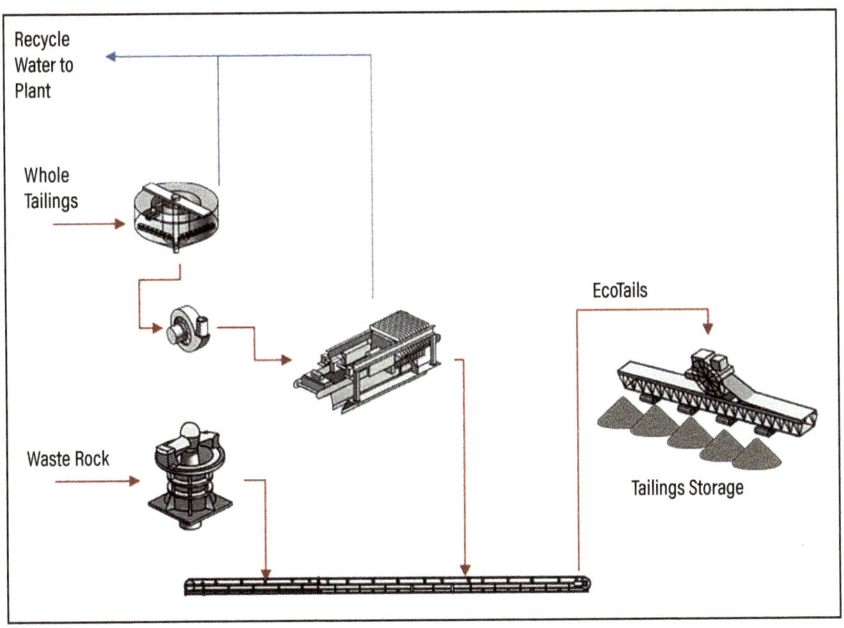

Courtesy of FLSmidth

Figure 6 How to make GeoWaste

use continues to increase as ore processing increases. Water shortage and water contamination continue to be issues for most mines. TMF failures are the most significant environmental liability for a mining project. Recovering water from mine tailings and disposing of the solids in a sustainable way are significant challenges for the mining industry. New processes, like EcoTails, and larger pressure filters, like a 5×3 filter press, have been developed to aid in addressing these challenges.

Courtesy of FLSmidth

Figure 7 GeoWaste

REFERENCES

Rahal, K. 2019. High efficiency filters for tailings applications. Presented at Tailings 2019, Santiago, Chile, July 10–12.

Wisdom, T. 2018. GeoWaste: Continuous co-mingled tailings for large scale mines. Presented at Paste 2018, Perth, Australia, April 11–13.

Wisdom, T. 2019. Maintaining high availability and low operational costs for filtered tailings facilities. Presented at Paste 2019, Cape Town, South Africa, May 8–10.

SESSION 5

The Importance of Culture in Safe Tailings Management

CHAIRS

Richard Davidson, AECOM
Tamara Johndrow, Freeport-McMoRan Inc.

PRESENTERS AND PANELISTS

Patrick Corser, Stantec
Richard Davidson, AECOM
Paula Doughty, Rio Tinto
Tamara Johndrow, Freeport-McMoRan Inc.
Kanyembo Katapa, Freeport-McMoRan Inc.
Andrew Watson, Stantec

NOTETAKERS

Samuel Abbaszadeh, Stantec
Peter Kowalewski, Tierra Group International, Ltd.

INTRODUCTION

The focus of this session is organizational culture and its importance in safe tailings management. An overview of the general process of cultural change within an organization and factors influencing the ultimate success in implementing that change were presented. The vital role of a company's safety culture in decreasing tailings storage facility (TSF) failures to a goal of zero was explored and focused on the following observations:

- Culture begins at the very top of an organization. Creating a strong safety culture with an environment of responsibility in a collaborative and supportive manner is critical.
- A high level of communication between qualified operational staff, subject matter experts, and multi-tiered reviewers helps illuminate

potential failure modes and actions that are required to improve resilience and reduce risk.

- Important traits include securing independent and competent technical and operational assessments asking engaged questions; making risk-informed decisions considering worst-case failure modes; searching for gaps in understanding, looking for hidden risks, and proactively mitigating them; and striving for continual improvement.

This session's presentations included cases studies, an analysis on the meaning of cultural change, and the importance of the role of engineer of record (EoR). Following the presentations, the panel participated in a guided question-and-answer session, which addressed the human elements involved with the safe design and management of TSFs.

PANEL DISCUSSION: THE IMPORTANCE OF CULTURE IN SAFE TAILINGS MANAGEMENT

There seems to be a "Goldilocks' point" that we are trying to reach, whereby we are confident enough to be transparent and yet not so risk averse that we do not actually move. We need to be able to create people who can confidently manage these facilities. How do you manage that?

The role of a review board (RB) is an advisory role: As engineers, you typically want to solve the problem, but it is important to let the project team do their job. You can provide advice and guidance to help them make wise choices. RBs do not have all details, whereas the people on-site, people building the structure, and the designers have those details. Together, you work as a team.

EoRs are involved in many RB meetings during which recommendations or suggestions are made and the owners sometimes think that they must follow through without question. Yet sometimes the guidance makes things less or more conservative. The obligation is on the EoR to digest that and decide what should be implemented. The EoR should be able to justify these decisions to the RB. It is important for there to be some back and forth between the EoR and the RB.

Owners appreciate having multiple pieces of advice. They first look to their EoR to dig into the design and analysis details and help make decisions. But owners are also very interested in hearing from their RBs, because owners know that they have such a wealth of experience working on many different projects. They often have seen projects over a long period and that wisdom is valuable. Owners update RBs on what they are doing; the RBs do not make decisions for owners but provide guidance on whether we are on the right path. Multiple points of view are invaluable.

Applying the right lessons from industry failures at other sites is beneficial. Design RBs can report directly to the operation. The RB is analogous to a second opinion from a doctor; if you are getting bad news or good news, you still want to make sure you have got somebody else looking at it. Sometimes RB members disagree with each other through constructive discussions of analysis approaches; those debates make for better and safer TSFs. It is important to have second opinions that help you to decide which risks are valid and which are not, and what needs to be done.

Is having a review board enough or can risk creep into our operation at many different levels?

The panelists all agreed that an RB is not enough; other measures also need to be implemented to identify risk. Other programs to identify risks, such as those used in FMI's Tailings Stewardship Program, are necessary, including risk assessments, identifying potential failure modes, challenging earlier design assumptions to assure they are still valid, and investigating facility conditions (cone penetrometer testing, drilling, other testing, etc.).

Projects require more than just design review. Regular reviews need to be completed to ensure that facilities are designed, constructed, operated, managed, and closed to their design intent throughout the mine life cycle. In addition, operations training, information sharing, visiting other operating sites to understand how they operate (looking at best practices) are all consequential. Finally, it is of utmost importance that the operations team understand the design, how the facility should be operated, and any impacts to safety.

The culture regarding transparency has dramatically changed: What more can we do to drive this change?

The panelists pointed out that investing in leadership training and increasing interaction with people leads to a better understanding of the situation. It is important to identify which lens you are looking through. We need to develop stronger relationships among the various operators, stakeholders, and regulators.

Other ways to drive change are to share best practices among mine operators. To raise the bar, industry-wide information sharing and transparency are needed. It is no longer good enough to just share information internally within the company.

Finally, the panelists agreed that although the Society for Mining, Metallurgy & Exploration (SME) Mine Tailings Symposium at the SME

2020 Annual Conference and Expo is a good start, they noted that generally only major miners participated. More needs to be done to engage mid- and junior-tier mining companies.

"A failure for one is a failure for all": How do we influence and help all operators (including the small and junior mining companies) adopt the best practices to help the whole industry succeed?

All companies can adopt the International Council on Mining and Metals (ICMM) guidelines as best practices. Updated tailings design guidance (not just regulations) should also be provided to all mining companies. Recognizing that culture manifests in our behavior and language, through co-locating and frequent sharing practices, people can improve and reinforce culture.

What can be done to facilitate and encourage more communication?

- Encourage and empower employees and contractors to speak up if unsafe conditions are identified. Follow through with actions.
- Collect monitoring data and make sure the right people are reviewing it. Technology now allows for real-time data reporting for review. Mobile apps allow operators and engineers to immediately report issues, allowing rapid engagement. Technology now allows remote or off-site EoRs to actively engage.
- RB meetings provide a great opportunity for younger staff to sit in and gain an understanding of the process. The meetings provide professional development as well as a chance for operations personnel to provide invaluable input.

How do we empower people to raise their voices and share what they see? How do we overcome "cultural issues" to instill culture?

It is important to encourage all levels of staff to point things out. For example, there are various levels of defense against TSF failure, including the EoR and the responsible dam engineer. Even before the responsible dam engineer, the initial layer comprises the operations personnel.

What do we want directors and CEOs of the corporations to do or say to reinforce this culture of taking care of the tailings storage facilities?

We want CEOs to trust us and empower us. We want them to hear us and have questions for us. If a CEO has questions, it means that person is interested. We want CEOs to understand the risks and the critical role of the EoR in the process, as well as the long-term relationship associated with being an EoR.

Owners and EoRs need to make investments in staff training and succession planning. Although difficult financial times might call for reductions in capital and personnel, we need CEOs to recognize that safety is the top priority.

What are the panelists' thoughts on the transfer of data or knowledge both internally and externally (to other companies)?

Information sharing is important. The panelists agreed that we are all in this together and need to share the knowledge that we have regarding safe tailings management across companies. For internal data and knowledge transfer, maintaining a hard-copy library of reports is invaluable. Transitioning information to a deputy EoR or the EoR's team is important.

What can attendees do to support the effort to change the culture?

The panelists offered the following ideas:

- Support young engineers and sponsor their development, such as attending conferences and participating in training courses and project field and office assignments.
- Diligent team members need to ask questions. Do not brush off concerns. Continue to share best practices.
- Companies should be willing to share information and experiences. Consultants need to capitalize on and present information at conferences.
- The career path for tailings should be seen as viable and exciting. We need to communicate with universities about tailings coursework and funded research. Most EoRs are civil and/or geotechnical engineers, so a specialization or certification in tailings would be a boon to the industry.

Some Observations on Cultural Change in Tailings Management

John Lupo
Newmont, Denver, Colorado, United States

INTRODUCTION

The title of Session 5 is "The Importance of Culture in Safe Tailings Management." Embedded in this title are two important, but interrelated aspects, namely, *culture* and *safety*. When we talk about safety, we really are talking about *risk* and, in this context, risk associated with tailings management. When we refer to culture, we are talking about people and their cultural traits, such as beliefs, values, and behavior. Although there are actually an infinite number of cultural traits we could discuss, for this paper, we are focusing on specific traits: risk and risk perception.

CULTURAL CHANGE

Merriam-Webster defines *cultural change* as "a modification of a society through innovation, invention, discovery, or contact with other societies" (Merriam-Webster, n.d.). Although this definition is not fully complete, it essentially means that a cultural change requires humans to change their habits (beliefs, behaviors, biases, etc.), usually in response to some external influence. In the context of tailings management, the catastrophic failures of tailings storage facilities (TSFs) over the past several years is the external influence that is driving cultural changes within the mining industry today.

At this point, it is important to emphasize that cultural change is solely a "human process." Often, cultural change within an organization or industry, is construed as adoption of new standards, procedures, processes, rules, and so forth. Although these are tools that can be used to influence a cultural change, these tools do not address the human process of change, which is the basis for cultural change. From my perspective, cultural change within tailings management must take a cue from social psychology to understand how humans interpret information and make decisions (e.g., risk identification and management). Understanding the influences on human decision-making is critical if there is to be change in a person's beliefs, behaviors, and biases, that is, cultural traits. In simple terms, it is easier to make something relatable if I know what influences a person's decision-making process and perception of risk.

DECISION-MAKING

The work of Weick and Daft (1984) and others suggests that humans interpret information and make decisions based on the following:

- **Self-esteem.** Confidence in yourself, personal identity, and what you think of yourself in relation to others affects the way you interpret information. Over-confidence in one's ability can lead to poor decision-making; conversely, a person with low self-esteem may be reluctant to make a decision. Recognizing how people perceive themselves within an organization and their own abilities is useful to judge decision-making. For example, a technician may feel that, because of his or her rank in an organization, he or she is not "qualified" to provide an opinion on management of a TSF or not responsible for implementation of a new TSF standard. Understanding how people perceive themselves and their roles in an organization is as important to tailings management as a new standard or procedure.
- **History.** Your past story, from where you were born and lived to what got you to where you are, all things in your personal history that have some influence in what you know and how you interpret the present. For example, in the context of TSFs, being born and raised in an arid environment may make it difficult to understand the challenges of tailings management in a tropical environment. We are all products of our own history, which makes us unique, but also influences our decisions and how we view risk. Personal history explains why two individuals can look at the same information yet arrive at drastically different opinions.
- **Social context.** Where you are in relation to others, what is happening around you, the nature of those around you, and the way others relate to the same information all influence the way you interpret information. Groupthink is a good example of social context, whereby your decisions are influenced by interpretations or decisions (right or wrong) of the group. Recognizing that some level of groupthink does occur in nearly all decisions is the first step in managing the process so that different interpretations and decisions can be heard and/or considered.
- **Confirming evidence.** We act something into belief, even creating a bias in our minds so that when something happens, it confirms the

belief. For example, there may be a belief that all upstream-raised TSFs fail, so that when a failure of an upstream-raised TSF occurs, it confirms this belief. Yet this belief ignores the context that there are numerous upstream-raised TSFs that are well managed and quite stable. It is difficult to challenge your own thought process, but doing so can lead to a more robust solution to a problem.

- **Cues and indicators.** What we see, hear, and feel does not necessarily carry information with it. We may recognize indicators and cues that give us information similar to things we have experienced before, but the information may be incomplete or biased. This explains how highly experienced individuals can still make poor decisions. All information is subjective and can be interpreted differently depending on the context and completeness of the information received.
- **Believability.** Our capacity to imagine is directly linked to not only what we believe but also to what we are willing to believe. Our ability to imagine extends or limits our ability to make sense of things. Believability is an important part of prediction and combines with past experience and cues to help us imagine what is possible. If we do not think something is possible, we do not plan for it and certainly cannot imagine the risks associated with it. This explains why, when assessing a low-likelihood/high-consequence event, we tend to focus on the low-likelihood (e.g., we do not believe it can happen) rather than the overall risk.
- **Flow.** The pace and speed of events affects the way we interpret them. Much of what we sense goes quickly to our subconscious and triggers a rapid intuitive response. Our intuition or gut feeling bypasses the need to process things step by step in a slow logical pattern. Our intuition gives us the fight-or-flight response we need in a crisis.

RISK PERCEPTION

From the preceding discussion, it is evident that people interpret information differently, so it is not surprising that risk perception differs as well. Herein lies the challenge for effecting a cultural change. An organization can adopt many approaches (e.g., styles) in an attempt to change the perception of risk and safety with management of TSFs; however, unless the organization embraces the concepts of learning, motivation, and considerations for how people learn, their focus remains fixated on systems, regulation, and the physicality of risk. The idea of risk ownership remains foreign to their organization. A comparison of various risk and safety styles is presented in Table 1. A perusal of this table suggests there are many methods or styles to address

Table 1 Comparison of risk and safety streams and style

	Orthodox Legal	Safety Science	Behavioral-Based Safety	Zero Harm	Process-Based Safety	People-Based Safety	Psychosocial Safety	Social Psychological Safety
View of humans	Human as servant	Human as object	Human as machine	Human as perfect	Human as part of system	Human as person	Well-being drives decision-making	Social relations drive decision-making
Focus	Rules, regulations, and standards	Method, order, and supposed logic	Rewards, monitoring, policing	Counting failure and compliance	Organization, systems, and glitches	Individuals, holistic safety	Well-being, mental health, and health	Social psychology, relationships, and neuropsychology
Language	Compliance, rules, punishment, control, consequence, systems, checklist, ALARP (as low as reasonably practicable)	Hazards, barrier, prevention, controls, consequence	Behavior, prevention, extrinsic, reward, punishment	"All accidents are preventable," aspiration, target, failure	Systemic error-failure, precedence, incubation, systems, methods	Human error, due diligence	Health, workplace, relationships, mental health, well-being, work-life balance	Risks, intrinsic motivation, heuristics, learning, mind, conversation
View of culture	Culture as systems	Culture as mechanics of systems	Culture as behavior	Culture-as-perfection-controls	Culture as organizational and leadership in systems	Culture as groups and leadership	Culture as holistic relationships	Culture as social construct
Strategy for change	Increased policing and systems	Increased barriers and controls	Increased surveillance and policing behaviors	Increased punishment and promotion of failure	Increased organizational intelligence	Increased focus on values	Increased focus on holistic relationships	Increased focus on social constructs and autonomy
Essential concepts	Hierarchy of control	Organizational systems	Observing and conditioning behaviors	Aspiration and target create reality	Reforming organizations	Tuning into people factors	Improving well-being and balance	Understanding and managing relationships and influences
Focus question	How can safety be organized?	What is the mechanics of safety?	How can people be controlled?	How many injuries would you like today?	How does the organization affect safety?	How can people minimize human error?	How can we keep the whole person well?	How do social arrangements affect decision-making?
Solutions	More engineering, technology, legislation, and regulation	Deconstruct mechanics, bowties, and barriers	Surveillance, training, positive and negative reinforcement	Counting failure, publish failure, preach aspirations	Improve organizations and leadership	Prioritize human factors	Enhance well-being and other aspects of safety follow	Learning and engagement through social relationships and attending

Source: Long 2017

risk, some of which are familiar to the reader. However, as the table illustrates, there is no single way or style for addressing risk. Why? Because the human aspect or element is present everywhere.

COMMUNICATION

Regardless of which style(s) is adopted from Table 1, communication is an essential part of any strategy to mitigate the misperception of risk. To alter the way people behave, we need to change their perception of risk. To effect that change, we must communicate. To communicate effectively, we need to understand an individual's viewpoint and then tailor our communication accordingly. We may need to address long-held preconceptions about what is a risk and what is not. Once people form an initial estimate of risk, they can be remarkably resistant to adjusting their perception, even when given new information (Breakwell 2007).

Communication can also reduce misperception caused by asymmetry of information. This asymmetry is created when someone is aware of a risk but does not share the information with other stakeholders within their organization. When two parties differ in their knowledge of a threat, their perception of risk is also likely to differ. It is difficult for users to care about a hazard if they do not even know it exists.

CONCLUSION

So how do we move forward on changing culture with respect to management of TSFs? To effect a cultural change, there must be recognition that change cannot rely only on standards, procedures, and processes. We, as an industry, need to consider how people interpret information that is used to form their perception of risk. Communication (face-to-face) is critical; otherwise, changes in perception of risk and culture are difficult to achieve.

In addition, when making key risk management decisions, a diversity of viewpoints must be included. Whenever possible, we should involve a broad cross section of individuals who represent groups across the organization. This diversity helps compensate for individual biases and helps align risk tolerance between the members and the organization. When decisions are to be made, it is important to ask penetrating, and perhaps uncomfortable, questions, intended to bring alternative viewpoints to the surface. These questions are designed to seek out the minority viewpoint that may be contrary to perceived wisdom. If the majority is telling the organization to turn right, are we missing something important that we would find out by turning left?

Cultural change can be achieved, but it requires a concerted effort that combines tools (standards, guidelines, training, etc.), focused and continued communications to align risk perception and risk tolerance, and adoption of performance measures to assure cultural traits are changing and are consistent with the organization's goals and objectives. Most of all, there needs to be a strong focus toward understanding human behavior through face-to-face discussions, group workshops, and continual feedback.

REFERENCES

Breakwell, G.M. 2007. *The Psychology of Risk*. Cambridge, UK: Cambridge University Press.

Long, R. 2017. Understanding the social psychology of risk. February 10. https://safetyrisk.net/understanding-the-social-psychology-of-risk-and-safety/. Accessed August 2020.

Merriam-Webster. n.d. s.v., "cultural change." https://unabridged.merriam-webster.com/unabridged/cultural%20change. Accessed June 2020.

Weick, K.E., and Daft, R.L. 1984. Toward a model of organizations as interpretation systems. *The Academy of Management Review* 9(2):284–295.

Integrating Company Culture and the Role of Engineer of Record

Patrick Corser
Stantec, Denver, Colorado, United States

INTRODUCTION

To understand how the role of company culture impacts the role of the engineer-of-record (EoR) and the safety of tailings dams, we first need to understand what we mean by *culture*. Then we can explore how different cultures—in a mining company and in an EoR company—can be integrated to improve the safety of tailings dams.

Definition of Culture

A culture is a way of life of a group of people—the behaviors, beliefs, values, and symbols that they accept, generally without thinking about them, and that are passed along by communication and interaction from one generation to the next. Every organization, including consulting engineers and mining companies, has a unique culture that has developed over many years and determines how they operate. Culture, therefore, greatly influences how structures like tailings storage facilities (TSFs) are designed, constructed, and operated.

Some aspects of organizational culture are visible on the surface, like the tip of an iceberg, while others are implicit and hidden from view, submerged within the organization (Figure 1). The implicit (below-water) attributes include beliefs, shared assumptions, values, traditions unwritten rules, perceptions, feelings, and stories. These are recognizable to anyone in the industry, but they are not the ones an organization advertises. The above-water attributes are the ones we hear about—the vision statements, strategy, and goals. The above-water attributes may describe *the way we say we get things done*, whereas the below-water attributes can be said to define *the way we really get things done*. Up to 90% of organizational attributes (Rick 2017) are below-water and when these are not aligned with the stated objectives of an organization, it cannot achieve its stated vision, strategy, and goals. This is

Source: Rick 2017, © Torben Rick

Figure 1 Iceberg to culture analogy

the challenge faced by the mining industry now: We know we must and have promised to improve how we manage tailings, and this requires a cultural revolution across the industry if we are to be successful. We know this can be done because the industry has made similar shifts before.

Safety Culture Success Story

As an example, we only need to look at environmental health and safety (EHS) performance. Over the last 15 to 20 years, the mining industry has led the way in aligning the culture of an entire industry, which resulted in significantly improving EHS performance at facilities across the industry and around the world. We regularly refer with pride to our safety culture and the reduction in workplace accidents and injuries (Figure 2). Consider for a moment the profound influence the industry-wide commitment to zero harm and zero injuries has had not only on the mining operations, but on the lives of workers, vendors, contractors, consultants, our families, and the communities we live in.

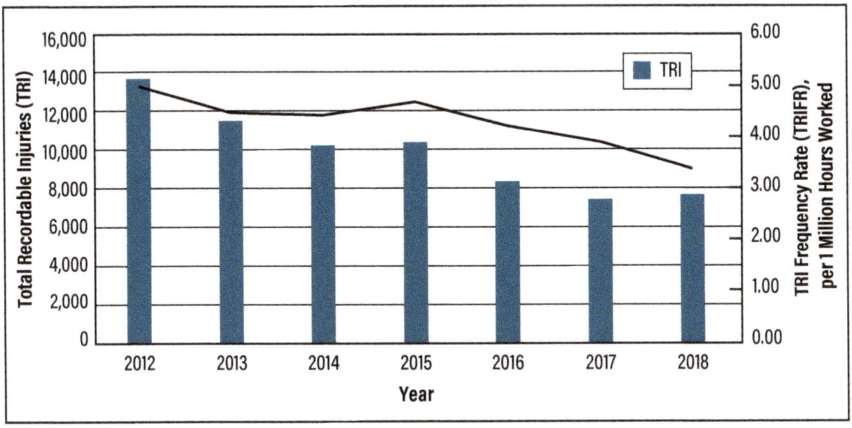

Source: ICMM 2018

Figure 2 ICMM total recordable injuries and frequency rate (2012-2018)

Integrating Company Cultures and Engineer-of-Record Cultures

Unfortunately, during that same period, the performance with respect to tailings disposal has not met commonly stated goals. Three catastrophic failures in the past three years have caused alarm and provoked stakeholders to question the industry's performance. We need to improve performance, and to do so, we need a cultural alignment around the design, construction, and operation of TSFs. This requires alignment within mining companies, and an alignment among the vendors, designers, contractors, and all their workers if we are to achieve results similar to the safety record of which we are proud. This cultural alignment is being done by some in the industry as evidenced by the papers presented in the Tailings Symposium as part of the Society for Mining, Metallurgy & Exploration (SME) 2020 conference. But this cultural alignment needs to be adopted across the entire industry to achieve the performance that we all expect.

Generally, the role of a mining company (at corporate and facility level) is to manage their resources and facilities (including the TSF) to produce minerals while protecting human health and the environment and respecting social commitments. The role of the EoR is to provide sound advice to the owner and assurance to other stakeholders that the TSF (dam and other components critical to performance of the TSF) has been designed and constructed, and is being operated in accordance with generally accepted standards.

The appointment of an EoR for a TSF has become widely accepted and is considered an industry best practice. However, for the relationship to work to the benefit of the TSF, the mining company, EoR, and their respective corporate cultures must be aligned (Figure 3).

Just saying "we are aligned" is not enough. Many in the industry are tackling this challenge, and this paper does not pretend to tell everyone's story. The following describes the effort a team at Stantec is making to play the part as engineers, designers of TSFs, and EoR for several facilities. We believe that if the entire industry embraces the culture needed to support the right behaviors, we can together achieve the common commitment to zero failures. This is our story.

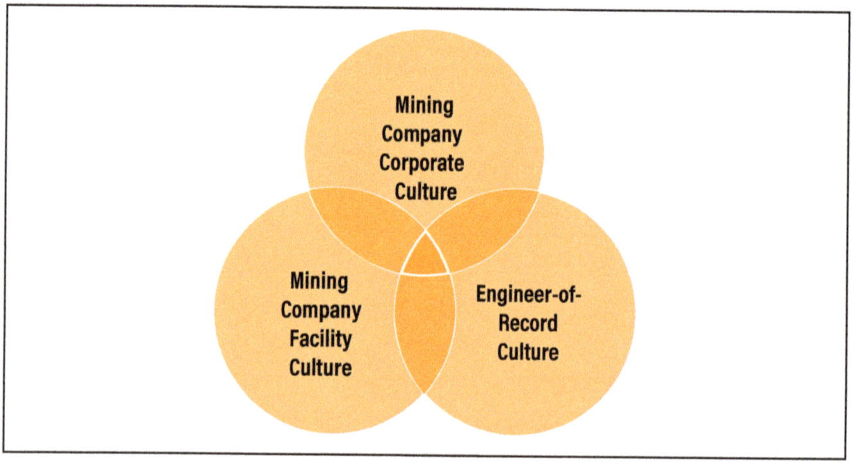

Figure 3 Integrated cultures to achieve industry-wide performance improvement

STANTEC'S ENGINEER-OF-RECORD CULTURE

Stantec seeks to extend a well-established dam engineering practice to address the demand for TSF engineering and EoR services. The market demand for review of the many TSFs causes us to accelerate and expand our efforts to prepare engineers. Usually, we meet the demand created by normal turnover and retirement through a steady process of experience-based training and increasing responsibility. However, an overall limit to the number of qualified EoR candidates in the industry makes recruiting difficult. In addition, the increasing demands made of the industry with respect to the role of the EoR requires careful selection of the persons assigned these responsibilities. Someone is not an EoR just because they say so.

Stantec Internal Engineer-of-Record Initiatives

We have pulled our tailings practitioners and those performing related engineering work together to establish a common sense of risk, discuss the necessary technical rigor, reinforce the need to observe and question, empower our professionals to speak up about what they see, and formulate a common approach.

The following is not a complete list of efforts that the entire industry must make. It is what we have embarked on so far and is shared with you in this paper in the hope that some of these measures can be of use to your organization and your circumstance.

- Established an EoR steering committee, composed of active senior EoRs and senior management
- Specified requirements for an EoR scope of work, establishing guidelines for client selection and specifying types of TSF designs where we consider assuming the role of EoR and conditions under which we would not accept the EoR role for an existing facility
- Supported senior management in a rigorous risk assessment process before new EoR assignments can be accepted and participated in periodic governance calls to make sure the EoR team has the support they need to fulfill their role
- Coordinated annual EoR summits hosted by the steering committee to review EoR work being globally performed, promoting Stantec best practices, capturing lessons learned, listening for industry trends, sharing client feedback, and discussing how our approach to tailings management might need to change
- Planned for and encouraged regular field inspections by the EoR team and the responsible dam engineer (RDE) team to understand construction

procedures and construction quality assurance (CQA) procedures, monitoring procedures, and observe first-hand the facility performance

Stantec Process for Engineer-of-Record Identification and Training

A shortage of tailings practitioners and qualified EoR staff exists throughout the industry. Therefore, we initiated a companywide weekly webinar series specifically focused on tailings, to attract and identify engineers with relevant skills and an interest in tailings. This self-selected pool of engineers at a variety of experience levels is where we are focusing our training and knowledge transfer efforts. Participation is voluntary, and we regularly attract more than 75 participants from more than five countries. We believe our future EoRs and deputy EoRs (DEoRs) are among them.

Stantec encourages the use of storytelling about successful projects, interesting and uncommon challenges, and novel designs. We also discuss projects or approaches that did not meet expectations, exploring how things might have been done better. Successful project descriptions are remembered, but honest exploration of challenging projects and mistakes are extraordinarily powerful learning opportunities.

Stantec has developed a framework for identifying potential EoR and DEoR candidates already on our staff. We have described the profile and experience we need to equip these individuals for the role, and we are working to provide experience, mentoring, and education as appropriate.

We continue to take our people to see projects, find them assignments that provide learning opportunities appropriate to their skill levels, and share our mistakes so we do not each have to learn the hard way. We are pressed to accelerate this process.

Like all of you, we also have multiple internal and external recruiters seeking experienced staff to grow our team. We have no illusion about how much work is required to mold this talented pool that includes young and old, experts and novices, old hands and new recruits into a cohesive team that embraces the integrity of the TSF. That is our cultural challenge.

CULTURE OF TAILINGS STORAGE FACLITY SAFETY

The performance of the TSF depends on the behavior of the team assigned to care for it. Our behavior is guided by our attitude, the conversations we have, and the instructions we receive.

It has been reported in industry surveys that most deficiencies at a tailings facility are not known to managers or supervisors but have been noted by frontline field staff (Figure 4). On the one hand, as discussed in the preceding section, we need to school and ready our people to prepare them for

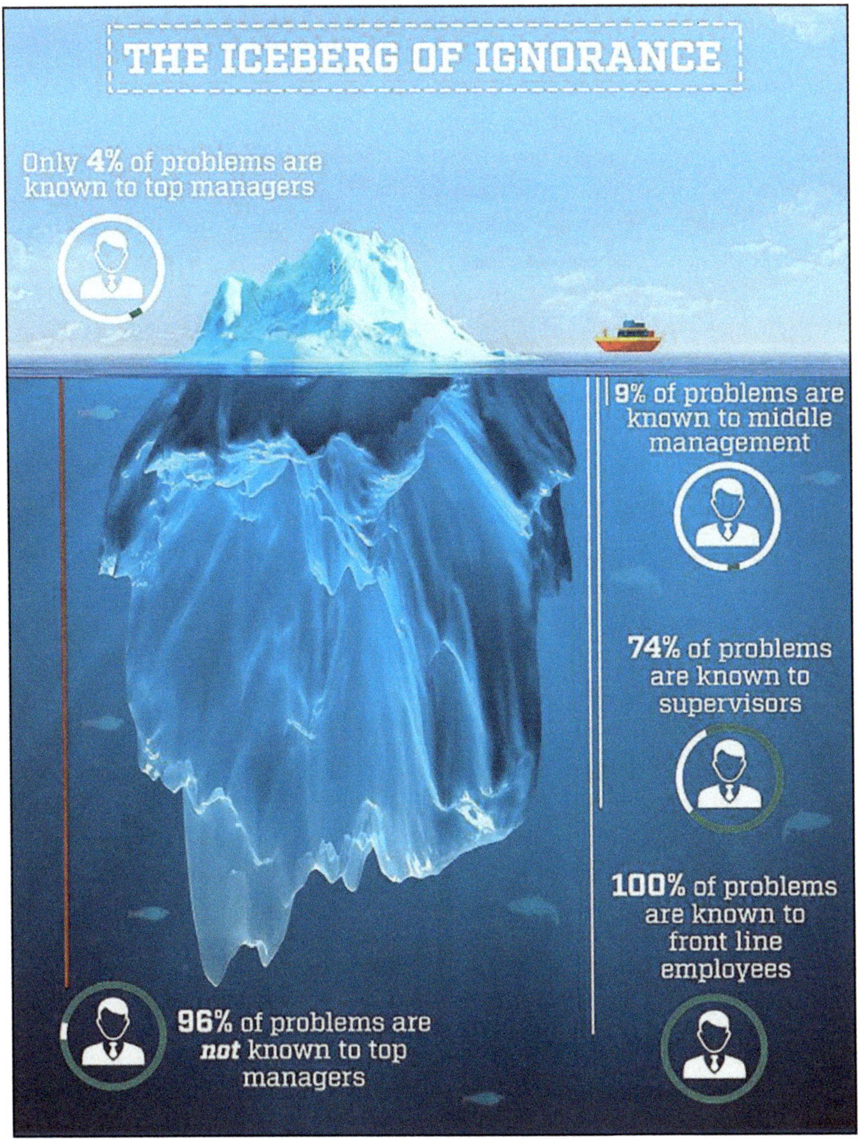

Source: Rick 2016, © Torben Rick

Figure 4 Iceberg of communication disconnects analogy

tomorrow. But in another sense, they know best what is *actually going on with the TSF today*, and we must listen to them.

Telling the team (and other stakeholders) that things are fine or will be fine is not the same as *listening* to those we have put in charge of the situation. This is the essence of the cultural gap discussed in the introduction. Yes, a strong TSF safety culture within the industry requires a dedicated group of EoR practitioners who share a common goal. And, ensuring the integrity of the TSF requires the collaboration, at every level, of every person in the supply chain and chain of command. If that sounds familiar, it is. That is precisely how we manage EHS. We are not safe because we have a safety officer, or because of the Mine Safety and Health Administration. Yes, we need competent and dedicated people to fill the EoR role, and we need everyone to fulfill the promise of their respective roles. The following are some elements we are working to make part of the Stantec TSF culture:

- **Values.** TSF performance is influenced by many internal and external stakeholders. Input from all stakeholders must be understood and evaluated as part of the whole facility assessment. This requires active engagement of the mining company teams, site operations teams, site environmental teams, EoR team, and regulatory representatives.
- **Structure.** Assignment of an external EoR and an RDE from the owners' organization and fostering that relationship is essential to curating the history and preserving the integrity of the TSF. Appointing deputies for these positions goes a long way to readying the next generation and preparing for the long-term care of the TSF. The EoR and RDE should have sponsorship from the very top of the organization, just as we have now for EHS. A RACI (which stands for responsible, accountable, consulted, and informed) matrix can be used to identify roles and channels of communication.
- **Norms.** Various industry bodies are specifying or proposing the type and extent of EoR engagement in planning, design, permitting, environmental monitoring, TSF CQA monitoring, dam performance monitoring, and emergency response planning. This dialogue must continue as we approach a common practice that serves the industry and our stakeholders.
- **Unwritten rules.** The EoR (and the CEO) can only see what they are shown and answer the questions they are asked. Some proven health, safety, and environment (HSE) practices, such as right-to-know, stop work authority, training, education, and dialogue, are needed so that the visible culture we speak about is infused into, embraced by, and become one with the below-the-water elements of our culture. The

unwritten rules by which we actually get things done reflect our real values and, when we get this right, enable zero TSF failures.

- **Stories.** The engagement of external review boards facilitates less biased review of facility conditions and allows some exchange of lessons learned at other facilities. We all have our stories. In HSE, we openly share our near misses, misfortunes, and lessons learned, and there is no shame in walking the path to zero harm. For tailings, we have not achieved that level of maturity, yet. A treasure trove of information inherent in our collective experience abounds—books to be written and stories to be told. However, they are constrained by confidentiality agreements and a reluctance to share unfavorable stories. Unless the industry agrees to share, we are doomed to repeat each other's mistakes.
- **Perceptions.** Historically, TSFs have been considered as a cost center, in effect, the tail end of the process. Now we see the TSF is also an essential part of the process, and the TSF outlives the mine. Neglected, the TSF can eat up all the profit ever generated and more. Failure of the TSF can cripple the mining company. We need to change our perceptions and priorities if we are to sway our stakeholders' perception of the industry.

CONCLUSION

Tailings management is serious business and getting it wrong has very serious consequences for our business, our industry, and the communities that support us. We know what a culture of safety means, and we have achieved some success with behavior-based loss-prevention initiatives. We know that the effort to have the culture embrace the initiative pays handsome dividends. At Stantec, we believe we can achieve a similar culture of tailings facility safety and know it is worth every penny the industry invests. This is our story about how we are shifting our culture. What is your story?

ACKNOWLEDGMENTS

I would like to acknowledge my colleagues, Andrew Watson, Resa Furey, and Amanda Adams, for their review and input on all aspects of this paper.

REFERENCES

ICMM (International Council on Mining and Metals). 2018. Benchmarking safety data: Progress of ICMM company members. www.icmm.com/safety-data-2018. Accessed August 2020.

Rick, T. 2016. Organizations rise or fall on their leadership—iceberg of ignorance. May 19. www.torbenrick.eu/t/r/aii. Accessed August 2020.

Rick, T. 2017. Important parts of the culture are essentially invisible. September 7. www.torbenrick.eu/t/r/efi. Accessed August 2020.

Rio Tinto Tailings Culture

Paula Doughty
Rio Tinto Kennecott Copper, Salt Lake City, Utah, United States

INTRODUCTION

Responsible use of the world's resources is key to a sustainable future for all. At Rio Tinto, we contribute to a more sustainable world through providing the materials essential for human progress in a responsible way. Part of that is how we manage the waste from mining operations. We manage the operation of tailings storage facilities (TSFs) at 26 sites worldwide and 12 closed and legacy sites. Most sites contain multiple storage facilities, and Rio Tinto has 96 tailings facilities, with a further 40 TSFs that are closed or under rehabilitation.

Our employees and the communities around our operations should expect that we understand our risks and have effective controls in place. For some time, we have had a robust approach to managing our sites. Since 2015, we have had a comprehensive safety standard covering both tailings and water storage facilities. Standards, however, are only effective if they are applied consistently and well. This standard is supported by three levels of assurance in our business.

INTEGRATED APPROACH TO MANAGING TAILINGS FACILITY RISK

We have a structured approach to managing tailings facility risks, underpinned by our safety standard. Figure 1 shows the three levels of governance and assurance that we apply to all facilities.

- The first level of assurance takes place at the asset itself with the main principles being effective facility design, comprehensive operational controls, monitoring, and regular reviews.
- The second level is assurance to the Rio Tinto standard through periodic business conformance audits and independent technical reviews. Business conformance audits are audits of an operation to assess conformance to the Rio Tinto health, safety, environment, and community's

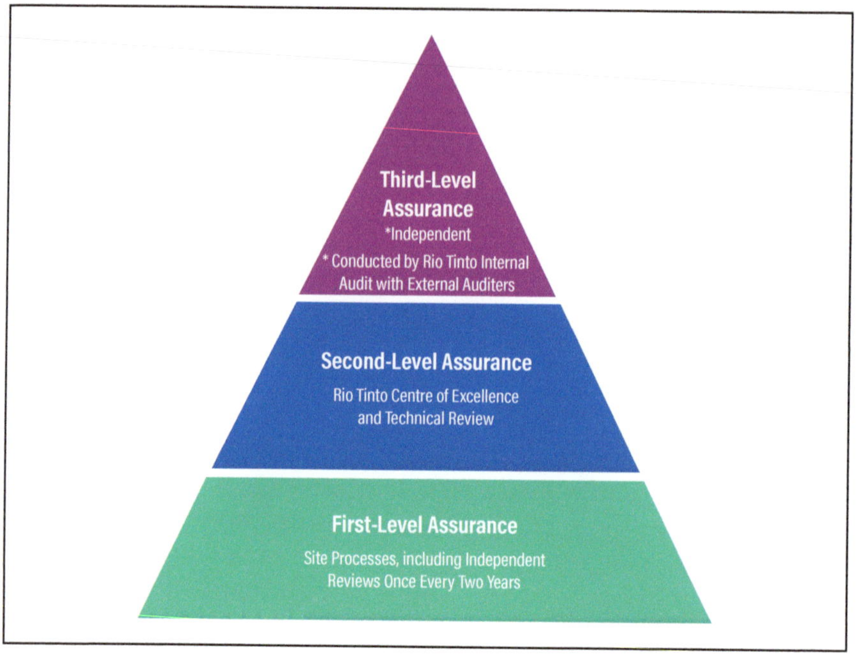

Courtesy of Rio Tinto

Figure 1 Levels of assurance

performance standards and management system according to established procedures and protocols.

- The third level of assurance is independent of site management and is normally conducted by third parties. The Rio Tinto Group internal auditors work with external auditors to provide confidence that the systems for risk management, internal control, and governance are adequate and effective.

All Rio Tinto–managed facilities, whether active or inactive, have an external engineer of record or design engineer.

GLOBAL SAFETY STANDARD

Although all of our sites identify and comply with laws, regulations, permits, licenses, and regulatory requirements, we went further with the introduction of a global standard for all tailings and water storage facilities in August 2015 to ensure that all of our managed facilities around the world are operated in accordance with this one global standard. The aim is to ensure that we protect the health and safety of people and the surrounding environment and preserve water in the best way we can everywhere we operate.

We joined the International Council on Mining and Metals (ICMM) Tailings Working Committee in 2016 to perform a tailings management review of member companies. Rio Tinto's work helped inform the ICMM Position Statement, including the six elements of tailings storage facility governance (ICMM 2016). Our standard is consistent with these six key elements:

1. **Accountability, responsibility, and competency.** Accountabilities, responsibilities, and associated competencies are defined to support appropriate identification and management of TSF risks.
2. **Planning and resourcing.** The financial and human resources needed to support continued TSF management and governance are maintained throughout a facility's life cycle.
3. **Risk management.** Risk management associated with TSFs includes risk identification, an appropriate control regime, and the verification of control performance.
4. **Change management.** Risks associated with potential changes are assessed, controlled, and communicated to avoid inadvertently compromising TSF integrity.
5. **Emergency preparedness and response.** Processes are in place to recognize and respond to impending failure of TSFs and mitigate the potential impacts arising from a potentially catastrophic failure.
6. **Review and assurance.** Internal and external review and assurance processes are in place so the controls for TSF risks can be comprehensively assessed and continually improved.

RECENT ENHANCEMENTS TO OUR TAILINGS MANAGEMENT APPROACH

Tailings management has been identified as a risk and classified in Rio Tinto's Group Risk Register since 2010. Since the establishment of our global safety standard for tailings in 2015, and in addition to our work as part of the ICMM working committee, we have made the following improvements to our approach:

- Developed a global procedure to mandate detailed control requirements to assist with the consistent implementation of the standard;
- Established the Surface Mining Center of Excellence Tailings team to provide technical expertise, ownership of the technical content of the global tailings' safety standard, and a second line of assurance in undertaking technical risk reviews of our facilities;
- Developed seven training modules on the leading practices for safe tailings management to provide specific training for the tailings dam operators on a regular basis;

- Appointed nominated managers and qualified site representatives for all TSFs. Nominated managers are accountable for their site's conformance to the Rio Tinto standard for the management of tailings and water storage facilities and qualified site representatives are accountable for the day-to-day operations and monitoring of tailings facilities;
- Review all designs and major studies for operational TSFs by our Surface Mining Center of Excellence Tailings team as well as by an independent third party or board; and
- Update our safety standard and procedure as required to ensure that we continue to learn from and embed any tailings audit outcomes.

FOCUS ON VALIDATION AND TRANSPARENCY

In February 2019, we established an internal task force reporting to the Center of Excellence to validate the data on our managed and non-managed tailings and water storage facilities, assess risk control effectiveness, and enhance technical assurance.

With the work of the task force, we are able to address questions, contribute to an industry response, and publicly disclose the hazard classifications for our tailings facilities and the current state of our facilities. We expect greater scrutiny of our activities, and we are very open to greater transparency.

EMERGENCY RESPONSE PLANNING AND COMMUNICATIONS

All of our operations have emergency response plans and follow business resilience and communications protocols. We perform a series of exercises and training initiatives, which are critical to test our plans in emergencies.

Although we have controls in place to manage the risks that tailings pose to people and the environment, we are not complacent. We have our processes assured by both internal and independent third parties. We know there is more for us to learn about best practices, so we partner with our peers to research and better understand them. Facilities can be safe with the application of appropriate design, planning, implementation, operational monitoring, and management standards. These and an active global tailings technical community allow Rio Tinto to continue to effectively manage the risks presented by these facilities.

REFERENCE

ICMM (International Council on Mining and Metals). 2016. *Position Statement on Preventing Catastrophic Failure of Tailings Storage Facilities.* London, UK: International Council on Mining and Metals.

Freeport-McMoRan's Tailings Stewardship Culture and Case Study

Kanyembo Katapa
Freeport-McMoRan Inc., Phoenix, Arizona, United States

Tamara Johndrow
Freeport-McMoRan Inc., Phoenix, Arizona, United States

Edwin Konrath III
Freeport-McMoRan Sierrita Inc., Green Valley, Arizona, United States

Christina Winckler
AECOM, Denver, Colorado, United States

INTRODUCTION

Freeport-McMoRan Inc. (FMI)'s subsidiaries own and operate 18 active tailings storage facilities (TSFs), 16 of which are located in the United States, and 2 of which are located in Peru. FMI's subsidiaries also manage 58 inactive or reclaimed TSFs in the United States. We fully support and implement the core elements of the position statement of the International Council on Mining and Metals (ICMM) on "preventing catastrophic failure of tailings storage facilities" (ICMM 2016), which are accountability, responsibility, and competency; planning and resourcing; risk management; change management; emergency preparedness and response; and review and assurance. We recognize that the failure of tailings and other impoundments at any of our mining operations could cause severe, and in some cases catastrophic, property and environmental damage and loss of life.

Our Tailings Management Stewardship Program was launched by corporate executive leadership in 2004 and has evolved over the last 15 years. Prior to 2004, the TSFs were managed at each mine site with the support of external engineers. In 2004, we formed the Tailings Stewardship Team (TST). The TST consisted of external engineering consultants and internal engineers from various mine sites who split their time between principal site duties and the TST. At inception, the TST, led by independent third-party engineers and reviewers, visited six to eight TSFs a year to review operational practices and recommend necessary changes. In 2008, the Technical Services Tailings, Crushed Leach, and Water (TS TCLW) group was formed at the corporate level with a mandate to provide portfolio-wide comprehensive support and resources for the TSFs. Today, through our Tailings Management System (TMS) led by the TS TCLW, we employ substantial engineering expertise, technological monitoring, and local and corporate management oversight to confirm that our TSFs are designed, built, operated, closed, and monitored to minimize risk to employees, neighboring communities, and the environment.

Our TMS defines the roles and responsibilities of internal and external experts who comprise our multi-tiered oversight. Oversight starts with the operational management and engineering teams at our sites. The TS TCLW at the corporate level serves as a dedicated technical resource for the site operators and engineers as well as an additional layer of oversight. TS TCLW does not have day-to-day operating responsibilities and exists solely to provide support and resources for our tailings dams. In addition to regular and frequent communication with site management, the corporate team communicates its activities and any findings directly to senior corporate management, which maintains active involvement and oversight with respect to our company's tailings management and stewardship.

The TST, comprising third-party engineers and internal experts, physically inspects our tailings facilities, reviews associated documents and monitoring data to identify potentially significant deficiencies, and recommends corrective actions. We have commissioned technical review boards (generally three- to four-person panels comprising internationally known experts, typically those with advanced degrees with more than 30 years of experience in their respective areas of specialization). These boards provide independent opinions and guidance on the physical integrity, safety, and performance of our TSFs and associated management systems.

Although we believe our tailings management and stewardship program is robust and mitigates risks, we continue to identify opportunities for continuous improvement in all aspects of our program. Our corporate culture is to communicate critical information upward so that important matters are identified, evaluated, and acted on. We have a strong commitment from our board of directors and executive management team to provide the necessary resources to enhance the structural stability and safe operations of our TSFs. FMI tailings management culture is founded on (1) operational engagement and open communication, (2) striving to identify and manage risks, (3) bias for action, and (4) continuous improvement. This paper elaborates on these four pillars and provides an example case study.

FREEPORT-McMoRan's CULTURE

FMI has instituted systems and procedures to ensure safe operation of our TSFs. Our employees are trained on the elements of our TMS. Over the years, we have also developed informal behaviors and beliefs that form our culture of tailings management. The culture is fostered through formal and informal interactions. The FMI culture comprises operational engagement and open communication; striving to identify and manage risks; bias for action; and continuous improvement.

Operational Engagement and Open Communication

Our TMS engages every level of the organization. Our multi-tiered review, support, and oversight groups include individuals as follows:

- Internal personnel
  - Site tailings operations team
  - Site TCLW team
  - TS TCLW point of contact and technical advisors
  - Site management
  - Site support departments
  - TST members (internal engineers)
  - FMI senior leadership team
- External personnel
  - Engineer of record (EoR)
  - TST (independent third-party reviewers [engineers])
  - Independent technical review board (TRB)

FMI encourages open bottom-up and top-down communication to drive action. For example, our frontline operators are empowered to communicate concerns directly to FMI management, and our EoRs routinely communicate with site teams and the TS TCLW team. Our TRBs have direct access to the chief operating officer for operating sites and direct access to the vice president of environmental services and sustainable development for inactive and closed sites, and members have a responsibility to report any concerns not being addressed by line management. Our culture of open communication allows us to quickly communicate potential issues to various levels within our organization and marshal resources as required to address them in a timely manner. FMI is committed to the safety and sustainability of communities in which it operates and maintains positive relationships with those communities. Each site communicates with community stakeholders through regular meetings during which company representatives transparently discuss operational activities, which encourages questions and an open dialogue. In addition, a dedicated community liaison officer facilitates a range of other ongoing interactions between operations personnel and community stakeholders. We stress to our employees, from corporate executive leadership to entry level, that bad news is never suppressed in our company.

Striving to Identify and Manage Risks

TSFs are complex structures with operating lives that can span decades. Turnover in management, engineering, and operations teams is inevitable over the life of a facility. In recognition of this, FMI dedicates substantial resources to continually inform and examine our site characterization models; address potential gaps in the engineering practice; identify, analyze, and manage risks through formal semi-quantitative risk assessments and as-needed special assessments; and apply lessons learned from past experiences, including industry failures.

In our efforts to continually inform and examine our site characterization model, FMI conducts periodic comprehensive geotechnical investigations to update the TSF material and pore pressure characterizations. These investigations are performed in addition to those addressing issues that may arise from normal TSF operations.

We conduct semi-quantitative risk analyses (SQRAs) to identify potential failure modes (PFMs) and determine if they are credible and if they are risk drivers. We conduct SQRAs in general conformance with dam safety industry standards. Our process includes site engineers and operators, corporate experts, the EoR, and an experienced external facilitator. PFMs are evaluated for normal loading conditions, earthquake loading up to design event, or flood loading up to design event or series of events. The results are used to determine if additional or alternative risk reduction measures are warranted.

We encourage our internal and external engineers to review past and present operating practices and apply lessons learned from past experiences and industry failures to our facilities. Our TRBs consider past failures and case histories to support their recommendations. Our sites and EoRs respond to these recommendations with additional investigations and analyses. We employ resources to remediate any findings of concern.

Bias for Action

FMI believes in providing the resources to ensure that it has the right people in the right technical positions. Everyone, regardless of position, is empowered to "stop the job" when they perceive an unsafe situation. This culture also extends to our TMS. We encourage our people to correct deficiencies wherever and whenever they are found.

We provide support structures that enable our employees to effectively perform their duties. We provide training, mentorship, and resources to both our tailings engineers and tailings operators. An example of training is the tailings short course for operators and supervisors. It is delivered over one to two days by the EoR or site senior engineer, typically in a classroom format. The content of the short course includes, among other things, tailings dam design, construction, and monitoring and inspection, with the goal of training operators on what "normal" looks like and examples of unusual field observation that should be reviewed with their supervisors and site engineers. Because operators spend most of their time at the TSF, likely they would be the first to observe unusual conditions. We also encourage our people to take pride in recognizing the importance of their role in dam safety. This sense of ownership drives proactive execution of their duties. Our employees know that they are supported in their actions to address deficiencies. We trust our employees to use the tools at their disposal (e.g., training, knowledge, peers, and internal and external technical experts) to make the right decisions. Through our multi-tiered oversight, those decisions are reviewed and refined if necessary. For example, an operator recently observed what he believed to be an unusual condition and initiated actions to stop deposition at one of our TSFs. The condition was investigated through our multi-tiered oversight and determined not to be a dam safety issue. The operator was commended for his actions.

When an issue is identified, FMI provides expertise, equipment, and resources from across the organization to help develop and implement solutions expeditiously. For example, when we decided to construct a buttress at Freeport-McMoRan Sierrita Inc. (FMSI), equipment operators from Freeport-McMoRan Morenci Inc. (FMMI), were temporarily reassigned to FMSI to expedite earthworks in preparation for the buttress.

Continuous Improvement

FMI's culture encourages employees to work individually and collectively to improve our TMS. Continuous improvement requires close coordination and communication between personnel involved with tailings management. Our TMS has three main components: personnel, programs and processes, and documentation:

1. **Personnel.** Multi-tiered oversight (e.g., site team, EoR, TS TCLW, other groups such as TST, and consultants) by diverse teams bring unique, experienced perspectives to tailings management. All participants are encouraged to provide feedback and identify opportunities for improvement. We hold regular collaboration meetings and an annual workshop where site teams share good tailings management practices and TS TCLW engineers share new initiatives, technology, and tools.

2. **Programs and processes.** We have developed and implemented numerous programs and processes as part of the TMS. The programs and processes are supported by documentation and are regularly reviewed and audited for effectiveness. The regular reviews and audits help generate feedback for the site tailings teams and TS TCLW to support updates and improvements.
3. **Documentation.** We prepare annual TSF reports that document construction, engineering, performance, and surveillance data for our TSFs and include a section for opportunities for improvement. Our TMS team members review activities for the previous year and investigate areas of improvement.

We recognize that the engineering practice continually evolves. FMI supports research and other initiatives to improve the state of the tailings engineering practice. For example, we are involved in the TAILLIQ Project (2018) with four Australian universities and other international mining companies. TAILLIQ is a university-industry research project whose objective is to help better understand and reduce the risk of mine tailings loss of containment caused by static liquefaction.

FMI recently identified a potential issue with an operational TSF at FMSI. A stability buttress was constructed to mitigate observed change in design conditions (Winckler et al. 2019). The project illustrates the pillars of our tailings management and stewardship culture.

CASE STUDY: SIERRITA BUTTRESS PROJECT

The Sierrita tailings impoundment (STI) is the only active TSF at FMSI. The STI has been an active upstream facility since 1970 and is raised approximately 6 to 8 ft per year. Initially, tailings were deposited behind a starter dam that was constructed along the eastern slope of the STI. The STI has an approximate height of 400 ft at the maximum section.

AECOM, with its legacy companies, has been involved with the STI since the 1970s. The current AECOM EoR assumed responsibility in 2012.

As part of FMI culture of identifying and managing risks, FMSI periodically performs comprehensive geotechnical investigations of the STI to update material and pore pressure characterization. FMSI's 2014 comprehensive geotechnical investigation included cone penetration test (CPT) soundings, boreholes, sample collection, laboratory testing, and geophysical suspension logging. The investigation also included interviews with past STI operators and review of historic aerial photos dating back to the construction of the STI starter dam in 1969. A lower resistance layer was encountered predominantly in the southern area of the STI. The lower resistance layer was indicated in CPT soundings by low tip resistance, low skin friction, and high dynamic pore pressures. The lower resistance layer is likely the result of the deposition of fine-grained tailings in topographic low spots and in borrow areas that were developed for the original starter dam construction. Historical photos of construction and the early operational period demonstrate this depositional pattern. Reviewing this history and supplementing available information with additional investigations and analysis was recommended by the TRB and was key to understanding the STI construction from the early operational period.

The 2014 stability analyses based on the 2014 material and pore pressure characterization met or exceeded the design factors of safety. The input parameters were based on the partially saturated shell using drained shear strength parameters and low pore pressures as was observed in 2014; the 2014 approach and results were presented to and endorsed by the TRB.

The STI is monitored by the site engineering and operations team. In mid-2017, FMSI observed increasing pore pressure in the southeast corner of the south dam. The increasing pore pressure presented a potential change to the 2014 site stability model and led the EoR to reevaluate the lower resistance layer. FMSI and the EoR initiated an investigation program that included review of STI surveillance data, field investigations, laboratory tests, and installation of new instruments in the southern corner of the STI. Stability analyses were performed for the southeast slope using undrained strength parameters. Analyses were also performed for the lower bound sensitivity case with the lower resistance layer modeled using fully softened shear strength. The results showed that a stabilizing buttress was required for the lower bound sensitivity case.

Consistent with our culture of open communication, FMI senior management and the TRB were informed throughout the process of the rising pore pressures, the subsequent geotechnical investigations and analyses, and the ensuing recommendation for a stabilizing buttress. Our commitment to zero catastrophic structural failures and bias for action resulted in the decision to construct a precautionary buttress in an expedited manner drawing on FMI resources and construction contractors. TS TCLW was engaged with FMSI to support the EoR's buttress design. FMI leadership was engaged to reassign equipment operators from FMMI and fast track contract and funding approvals and construction contractor boarding.

Prior to construction initiation, FMSI engaged local stakeholders to inform them about the precautionary buttress project and the tentative schedule. Communications with state and local stakeholders emphasized the commitment to safe tailings facilities. FMSI management met with employees and sent communications outlining the STI site conditions and FMSI's associated action plan.

A buttress was designed in two phases: phase 1 for STI elevation 3,510 ft amsl (above mean sea level) (up to mid-2020) and phase 2 for STI elevation 3,585 ft amsl (approximately year 2029). Stability analyses demonstrate factors of safety for the undrained and post-earthquake loading conditions that exceed the minimum required values for the phases 1 and 2 buttress layouts. These buttresses were constructed before the STI reached the targeted elevations. The phase 1 buttress was completed in March 2019, and the phase 2 buttress was completed in September 2019.

The observational method is being implemented to monitor pore pressures, saturation within the STI tailings, and potential for any movement of the STI embankment. Future geotechnical investigations will periodically occur as standard practice. An updated comprehensive investigation and analysis is ongoing in 2020. If any unexpected changes in behavior are observed, the proposed buttress layout would be re-evaluated and modified, if required.

CONCLUSION

FMI's tailings stewardship culture is founded on operational engagement and open communication, striving to identify and manage risks, bias for action, and continuous improvement. FMI's stewardship culture has been fostered and encouraged over the past 15 years by interactions between executive leadership, TS TCLW corporate group, competent engineers and operators at our sites, our long-term partnerships with our EoRs, and many weeks of inspections, reviews, and engaging conversations we have received from our TST and our TRB members. Just as we organically developed our culture and programs to date, we will continue to evolve in the coming years as broader industry evolves.

REFERENCES

ICMM (International Council on Mining and Metals). 2016. *Position Statement on Preventing Catastrophic Failure of Tailings Storage Facilities*. London, UK: International Council on Mining and Metals.

TAILLIQ. 2018. Evaluating potential static liquefaction of tailings to prevent failures. TAILLIQ blog post. November. www.tailliq.com/.

Winckler, C., Katapa, K., Konrath, E., et al. 2019. Stability buttress to mitigate observed change in design conditions. In *Proceedings of Tailings and Mine Waste 2019*. Edited by J. Goodwill, D. van Zyl, and M. Davies. Vancouver: University of British Columbia. pp. 289–318.

Index

Note: *f.* indicates figure; *t.* indicates table

S

T

U

W